Richard Mackay is a Principal Environmental Planner at Mott MacDonald, specializing in ecology. He has a wide experience of wildlife surveys and environmental impact assessment, as well as formulating environmental policies for major companies. He is co-author of the World Health Organization's award-winning *Inheriting the World: The Atlas of Children's Health and the Environment.*

Praise for previous editions:

"Highly engaging pictures, maps and graphics that bring immediately home the ever-increasing crisis of extinction." *The Ecologist*

"The first fully illustrated and comprehensive guide to the world's endangered plants and animals." *The Bookseller*

"A pleasure to read...fantastic use of colour and photography." *What on Earth*

"A must for budding botanists and ecologists." *Rocky Road*

"Quite simply, every school and institution library should get a copy." *British Ecological Society*

"A compact, clear and informative book that should have a place on the bookshelf of all readers interested in nature conservation." *Entomofauna*

In the same series:

"Unique and uniquely beautiful. . . . A single map here tells us more about the world today than a dozen abstracts or scholarly tomes." *Los Angeles Times*

"A striking new approach to cartography. . . . No-one wishing to keep a grip on the reality of the world should be without these books." *International Herald Tribune*

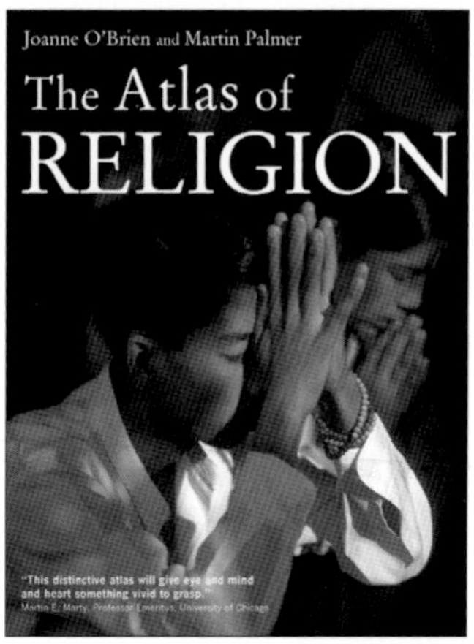

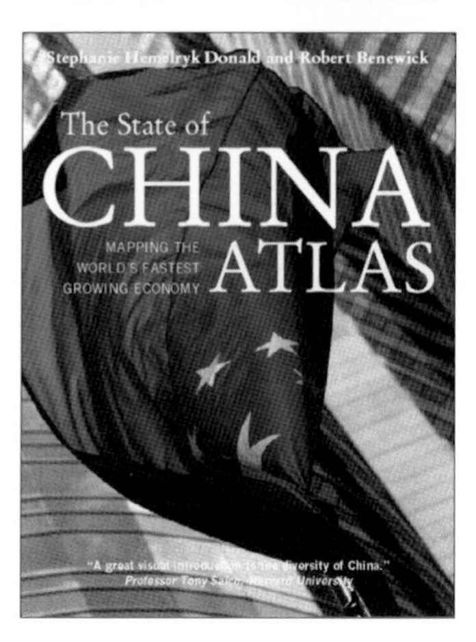

THE ATLAS OF ENDANGERED SPECIES

Revised and Updated

Richard Mackay

UNIVERSITY OF CALIFORNIA PRESS

Berkeley *Los Angeles*

University of California Press, one of the most distinguished university presses in the United States, enriches lives around the world by advancing scholarship in the humanities, social sciences, and natural sciences. Its activities are supported by the UC Press Foundation and by philanthropic contributions from individuals and institutions.
For more information, visit www.ucpress.edu.

University of California Press
Berkeley and Los Angeles, California

Cataloging-in-publication data for this title
is on file with the Library of Congress.

ISBN 978-0-520-25862-4 (pbk. : alk. paper)

Produced for the University of California Press by
Myriad Editions
Brighton, UK
www.MyriadEditions.com

Edited and co-ordinated for Myriad Editions by
Jannet King and Candida Lacey
Design and graphics by Corinne Pearlman and Isabelle Lewis
Maps created by Isabelle Lewis
Additional research by Jannet King

Printed on paper produced from sustainable sources.
Printed and bound in Hong Kong through Lion Production
under the supervision of Bob Cassels, The Hanway Press, London.

15 14 13 12 11 10 09 08
10 9 8 7 6 5 4 3 2 1

Contents

Part 5 Endangered Birds

Part 6 Issues of Conservation

Part 7 World Tables

Introduction

EVOLUTIONARY ADAPTATION HAS GIVEN RISE to a cornucopia of life. Of all the species that have ever lived, only a tiny fraction is alive today. Spasms in Earth's history, caused by volcanoes erupting and meteorites impacting, wiped out entire groups of organisms, and gradual climatic changes have eliminated others. The evolutionary strategies of some species were just not good enough and they succumbed to competitors and predators. The demise of the dinosaurs is a humbling tale. Their awesome size and power did not assure their survival. Quite the reverse: it left them slow to adapt to a changing environment and more vulnerable than mammals to catastrophic events.

As humans, an appreciation of our evolutionary origins has recently been complemented by a growing awareness of the genetic similarities we share with other living-creatures, particularly chimpanzees and bonobos. The Australopithecines, Homo ergaster and our other ancestors took tools, organization and culture well beyond the level of sophistication of the primates. Although we are of their line, the biological changes that accompanied their advancement are sufficient for us to consider ourselves a different "chrono-species".

Some parts of the world are more ecologically interesting than others. Habitats such as coral reefs and rainforests contain a great variety of species. Some intrigue us because, like the Arctic and Antarctic, their wildlife is peculiar and we have not yet heavily disturbed them. Species on islands such as Madagascar developed in geographical isolation, giving rise to unique forms. Australia has the dubious distinction of suffering the most mammal extinctions. Many other regions and ecosystems could have been included here but for the lack of space.

A small fraction – probably fewer than 10 percent – of living species have been identified and classified. The remaining 90 percent are mostly invertebrates. Consequently, an anomaly arises in that the most intensely researched countries show up as having the highest number of threatened species. The World Conservation Monitoring Centre, now a division of the United Nations Environment Programme, is the global data organization for conservation. IUCN is the umbrella group for conservation bodies around the world. It has promulgated a system for assessing whether a species is threatened and, if so, the severity of the risk. The definitive work is the organization's publication: the *IUCN Red List of Threatened Species.*

Representing the number of species threatened endorses the notion of a species as a closed gene-pool. This presents conceptual difficulties, such as in defining species among organisms that reproduce asexually, for example in animals that bud and in plants that spread by growth of horizontal stems. Despite its flaws, however, the species concept is useful for conservation. Subspecies and isolated populations of species are generally not counted in this book, although they are sometimes described in the text.

The distribution of living things does not respect political boundaries. However, decisions affecting conservation are still usually made by states, and their borders provide a convenient means of dividing the land surface. For groups of animals and plants, the level of threat has been represented in this atlas by the number of threatened species occurring in each country. This inevitably gives the impression that larger countries have greater conservation problems than smaller ones. To redress the balance, Part 6 Issues of Conservation highlights areas with the greatest biodiversity and those with the most endemic species.

Why should conserving biodiversity matter? Many people feel that we are stewards of the planet and that to annihilate species as we are doing is immoral. Quite apart from spiritual and aesthetic loss, each extinction also represents the loss of irreplaceable genetic information, including beneficial characteristics that could have been bred into domestic crops and livestock. Chemicals produced naturally are also used to develop pharmaceuticals. When "keystone" species become extinct the balance of whole ecosystems is threatened.

Global warming is one of a number of threats outlined in the *Millennium Ecosystem Assessment*, completed in March 2005. This was the first comprehensive global evaluation of the world's major ecosystems. The assessment found that 60 percent of ecosystem services (the benefits people obtain from ecosystems, such as fisheries and prevention of soil erosion) are being degraded or used unsustainably. Loss of species was identified as a major factor affecting both the resilience of natural services and less tangible spiritual or cultural values. The consequences for human well-being are severe, but by protecting ecosystem services in agriculture, industry and in our lifestyles, we can halt the decline.

Agreement between states can secure the conservation of common resources, such as ocean fisheries. It can also suppress damaging wildlife trade, driven by our fascination with the rare and exotic and the purported medicinal properties of some plants and animals. The Convention on International Trade in Endangered Species of Wild Fauna and Flora (CITES) was adopted in 1973. More recent treaties have addressed conservation within national jurisdictions. In 1992 the United Nations Conference on Environment and Development in Rio de Janeiro (the "Rio Summit") adopted the Convention on Biological Diversity, the Convention on Climate Change and Forest Principles. All are relevant to threatened species. Although the Convention on Biological Diversity restated the principle of national sovereignty, it also placed a duty on states to conserve biodiversity by requiring that environmental impact assessments be carried out for infrastructure projects, for example.

The size of the human population is of major consequence in conserving biodiversity. Fewer people would allow us to leave habitats intact or to exploit them sustainably. Of course the environmental impact, or "footprint", of individual countries differs. The USA, with four percent of the world's population, produces a quarter of the world's greenhouse-gas emissions. However, the USA and other industrialized countries also develop most of the world's clean technologies.

Pollution is a global issue. Carbon dioxide warms the whole planet, not just your back yard. The impact of global warming on threatened species will be profound. Polar bears will be stranded and starve as ice retreats. Plants with a limited distribution may not survive rising temperatures. Coasts and estuaries will be inundated. The world will lose species that have evolved over billions of years to form communities and habitats that without human interference would still exist in exquisite equilibrium.

The Convention on Climate Change went to the very heart of economic systems. An accord was signed in Kyoto in 1997, outlining binding limits on the global warming contributions of developed countries. The extent to which the transfer of clean technology to developing countries and "carbon sinks" could be debited from national emissions was contentious. In November 2000, talks in The Hague, Netherlands on the detail of the Kyoto agreement foundered on US intransigence, but in July 2001 a compromise agreement was reached by 180 countries (excluding the USA). Following ratification by Russia, the Kyoto Protocol finally came into force on 16 February 2005. Talks in Bali in 2007 aimed to map out a framework for concerted international action on global warming to succeed the Kyoto Protocol, which expires in 2012. It is not yet clear whether the successor treaty will secure deep cuts in emissions by developed countries and implement targets for developing countries, which – particularly with the industrialization of China and India – are becoming significant emitters of greenhouse gases. Both are required if catastrophic climate change is to be averted.

RICHARD MACKAY
Cambridge, Summer 2008

ACKNOWLEDGEMENTS

My thanks to Caroline Pollock at the IUCN for providing data, to Patricia Patton at WWF-UK and Alex Solyom at WWF-US, who kindly supplied many of the photographs included in this book, and to the photographers who have generously given us permission to use their images (see below). I would also like to thank the Species Survival Commission of IUCN and Birdlife International, who were both helpful in clarifying data.

I am very grateful to the team at Myriad Editions – Jannet King, Candida Lacey, Isabelle Lewis and Corinne Pearlman – who worked tirelessly to ensure the rigor and visual impact of the book, and especially for Jannet's help researching the new data.

Another sincere acknowledgement is to my parents Judith and John, who gave their wholehearted support.

The author and publishers are grateful to WWF in the USA and UK for permission to use a selection of images from the WWF photo archives. Known worldwide by its Panda logo, WWF is working around the globe to protect endangered species and to save endangered spaces. To learn more about WWF and its programs, visit the organization's portal website at www.wwf.org. WWF assumes no responsibility for the accuracy of the text, maps or charts in the atlas. In addition to the WWF photographs, we are grateful for permission to use the following on pp.10–11: Giant tortoise: Neil Morrison WWF-UK. 16–17: Neanderthal skull: Patrick McDonnell/www.medicalillustration.net. 18–19: Turtle: Asther Lau/www.iStockphoto.com. 20–21: Red Sea coral: Charles Hood/WWF-UK. 22–23: Mahogany tree, Brazil: Mark Edwards/Still Pictures/WWF-UK; Tropical forest: WWF-UK; Rainforest destruction, Indonesia: Mauri Rautkari/WWF-UK. 24–25: Legal logging, Russia: UPM; Tree-planting, China: Asian-info. 26–27: Asia's high steppes: iStockphoto/Robert Churchill; great Indian bustard: Raja Purohit. 28–29: Mangroves: Edward Parker/WWF-UK; Water hyacinth: USDA. 30–31: Red Sea coral: Charles Hood/WWF-UK. 32–33: Orange roughy: Greenpeace/Duncan; Coral: NOAA/MBARI. 34–35: Gentoo penguin, Antarctica: iStockphoto/Jeff Goldman. 36–37: Polar bears: Neil Morrison/WWF-UK. 38–39: Macaroni penguins: Mary Rae/WWF-UK; Rockhopper penguin: David Lawson/WWF-UK. 40-41: Gouldian finch: Australian Wildlife Conservancy. 42–43: Sea lions: iStockphoto/Carrie Winegarden; Muriqui: Luiz Claudio Marigo/www.omuriqui.hpg.com.br; Unofficial roads: iStockphoto/Joseph Luoman; Pantantal: iStockphoto/Torsten Karock. 44–45: Galapagos penguin: iStockphoto/javaman3; Steven Morello/WWF-US; Lava gull: Gary Feldman; Giant tortoise: Neil Morrison/WWF-UK. 46–47: Madagascar rosy periwinkle: David R. Parks/www.mobot.org; Fish eagle: Greg Lasley; Golden bamboo lemur: endangeredcreatures.net. 48–49: Cheetah: Chris Harvey/WWF-UK. 50–51: Drill monkey: D. White/WWF-UK; Alaotra gentle lemur: David Lawson/WWF-UK; Orang-utan: Russel Mittermeier/WWF-US; Gorilla: Rick Weyerhaeuser/WWF-US. 52–53: Florida panther: David Maehr/Conservation Biology, University of Kentucky: www.fl-panther.com; Bengal tiger: David Lawson/WWF-UK; Cheetah: Chris Harvey/WWF-UK. 54–55: Bison: USFWS/WWF-US; Przewalski's horse: John De Meij/Foundation for the Preservation and Protection of the Przewalski Horse; Pygmy hog: Durrell Wildlife; Hunter's hartebeest: Gretchen Goodner/WWF-US. 56–57: Indian elephant: David Lawson/WWF-UK; Javan rhino: Thinkquest; Indian rhino: Bruce Bunting/WWF-US. 58–59: Brown bear: Mary Rae/WWF-UK; Giant panda: Edward Mendell/WWF-UK; Asiatic black bear: Ian Ledgerwood/WWF-UK; Spectacled bear: David Lawson/WWF-UK. 60–61: Beaver: David Lawson/WWF-UK. 64–65: Blue whale: Paul Coppi/WWF-UK. 66–67: Panamanian golden frog: www.messiah.edu; Marine iguana: T. P. Littlejohns/WWF-UK;. Asian three-striped box turtle: Kurt Buhlmann/Conservation International. 68–69: Harlequin ladybird: Wild About Britain; Monarch butterfly: WWF-US. 70–71: Aquaculture: Edward Parker/WWF-UK. 72–73: Lady's slipper orchid: E. Lister/WWF-UK; Wild bluebells: Jannet King; Caucasus: Cathy Ratcliff/WWF-UK; Mandrinette: Wendy Strahm/IUCN Red List Programme; Bastard quiver tree: Craig Hilton-Taylor/IUCN Red List Programme. 74–75: Short-tailed albatross: Hiroshi Hagesawa/Toho University. 76–77: Kiwi hen: Storm Stanley/WWF-UK. 78–79: Great Philippine eagle: www.philipineeagle.org. 80–81: Hyacinth macaw: Edward Parker/WWF-UK; Female Mauritius parakeet: Anne Lee/WWF-UK; Yellow-eared conure: Paul Salaman/Proyecto Ognorhynchus. 82–83: Short-tailed albatross: Hiroshi Hagesawa/Toho University. 84-85: Turtle Dove and Steppe Eagle: Sergey Dereliev/UNEP/AEWA. 86–87: Cloud forest, Costa Rica: iStockphoto/Francisco Romero. 88–89: Chimpanzee: David Lawson/WWF-UK; Red ruffed lemur: David Lawson/WWF-UK; Galapagos tortoise: Charles Hood/WWF-UK. 90–91: Everglades: iStockphoto/Harry Thomas; Epiphytes: Russel Mittermeier/WWF-US; Rainforest, Sumatra: Mauri Rautkari/WWF-UK. 92–93: Sequoia: iStockphoto/Tobia Peciva; Dinaric Arc: iStockphoto/Ralf Hirsch; Pygmy hippo: David Lawson/WWF-UK; Golden lion tamarin: David Lawson/WWF-UK. 94–95: Alaotran gentle lemur: James Morgan/Durrell Wildlife; Mallorcan midwife toad: Gerardo Garcia/Durrell Wildlife; Mauritius kestrel: Carl Jones/Durrell Wildlife; Black lion tamarin: Mark Pigeon/Durrell Wildlife. 96–97: Japanese knotweed: invasive.org; Svalbard Global Seed Vault: Mari Tefre/Global Crop Diversity Trust; Magnolia: Royal Horticultural Society; Golden pagoda: Craig Hilton-Taylor/IUCN Red List Programme;. 98–99: Angler sattelschwein/FAO; Red Maasai sheep/FAO. 100–01: African gray: Jannet King; 104–05: Monarch butterfly (photo montage): iStockphoto/Cathy Keifer.

Terrain bases on pages 26–27, 38–39, 43 and 92–93 were prepared using MAPS IN MINUTESTM © RH Publications (1999); pages 18–19, 32–33, 40–41, 46–47 use Mountain High maps © Myriad Editions; page 37 uses a base map created by Hugo Ahlenius and reproduced by kind permission of UNEP/GRID-Arendal. Available at: http://maps.grida.no

1 Extinction is Forever

"Pragmatic self-interest alone should teach us that
we must change before nature exacts inevitable revenge."

– David Watson,
author of *How Deep is Deep Ecology*

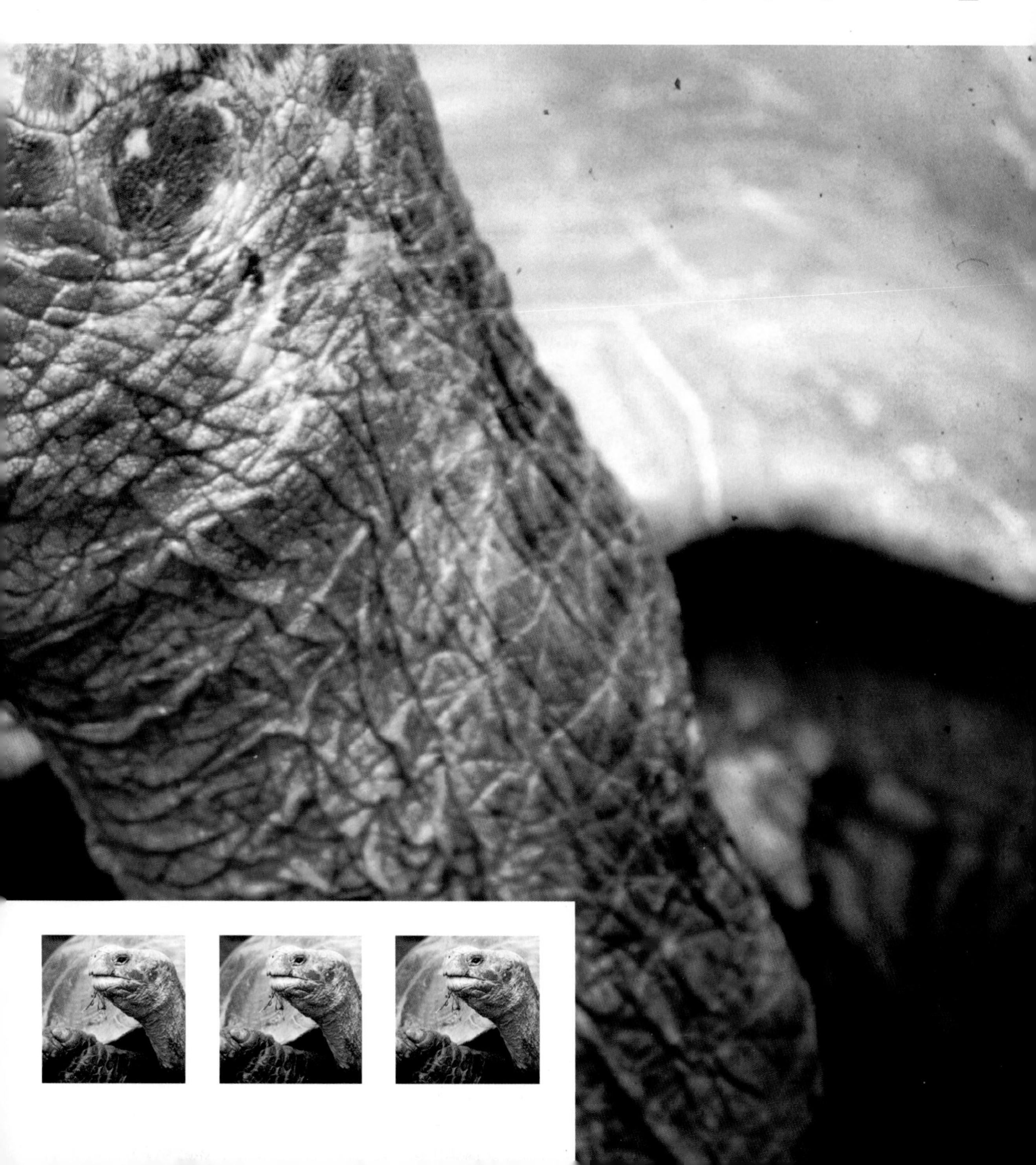

EVOLUTION

In 1831 the British naturalist Charles Darwin took a fateful voyage on the survey ship HMS *Beagle*. Over five years he studied the geology and wildlife of the lands the *Beagle* visited. He found fossils of species long extinct and wondered how new species had replaced them. His observations led him to deduce that the difference between species was an outcome of natural processes. Darwin's thesis culminated in the publication in 1859 of *On the Origin of Species*. Here he challenged the prevailing orthodoxy that the diversity of life was the product of supernatural design.

By Darwin's day the concept of a "species" was widely understood as a group of organisms that can breed with each other but cannot interbreed with another species. The Swedish botanist Karl von Linné (who became known by his Latin name of Carolus Linnaeus) had developed a system in the mid-18th century that classified organisms according to their similarities, placing them in a hierarchy that, at the highest level, makes the basic distinction between animals and plants. His approach forms the basis of modern-day taxonomic diagrams (see opposite).

The difference between the approach of Linnaeus and Darwin was that while Linnaeus saw his work as "mapping" the world God had created, Darwin was interested in explaining how living creatures evolved and continued to develop. Darwin observed that there is variation within a species and that some characteristics are more beneficial than others. He knew that the characteristics of an adult organism depend on its parents and the environment where it lives. As life is a competition for scarce resources and a test of resilience in the face of harsh conditions, organisms with characteristics most likely to allow them to survive will breed most successfully. Those characteristics will then become more prevalent within the species as it adapts – or evolves – to its changing circumstances. When events such as a rising sea level or the movement of the Earth's continents splits a population into two, the new populations cannot meet to interbreed. Over time, separate evolution will cause changes in each group to the point where they can no longer successfully reproduce with each other – a new species has evolved.

A taxonomic diagram, such as the simplified version opposite, relates species by their evolutionary history. Species of the same order are considered to share a more recent common ancestor than species of the same phylum. Different species with similar features are grouped into families. Similar families belong to an order, related orders to a class and classes to a phylum, each of which belongs to one of the kingdoms. There are, however, different criteria for assigning organisms to different groups, and continuing debates as how the divisions should be made, even at the highest level of the kingdom.

Darwin's theory of evolution is itself constantly being reviewed. In the 20th century scientists studying life at the molecular level recognized that natural selection occurs only indirectly between whole organisms (animals and plants). At a more fundamental level it is occurring between genes. This explains behaviors such as altruism: for example, ants will sacrifice themselves to defend their nests because what is important is the survival of the genes shared by related ants rather than survival of the individual.

EVOLUTION SIMPLIFIED

How a single species can evolve into many species

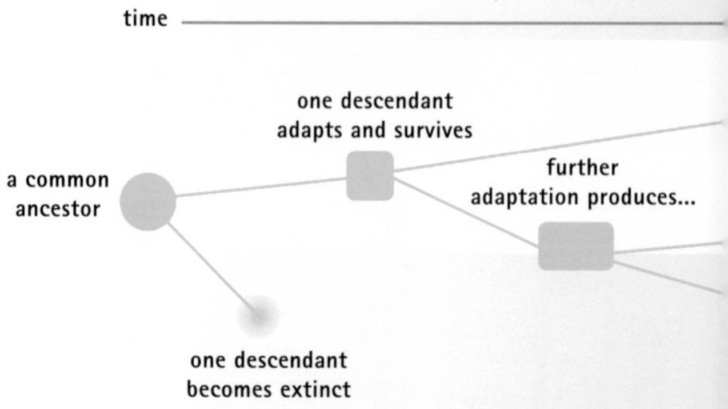

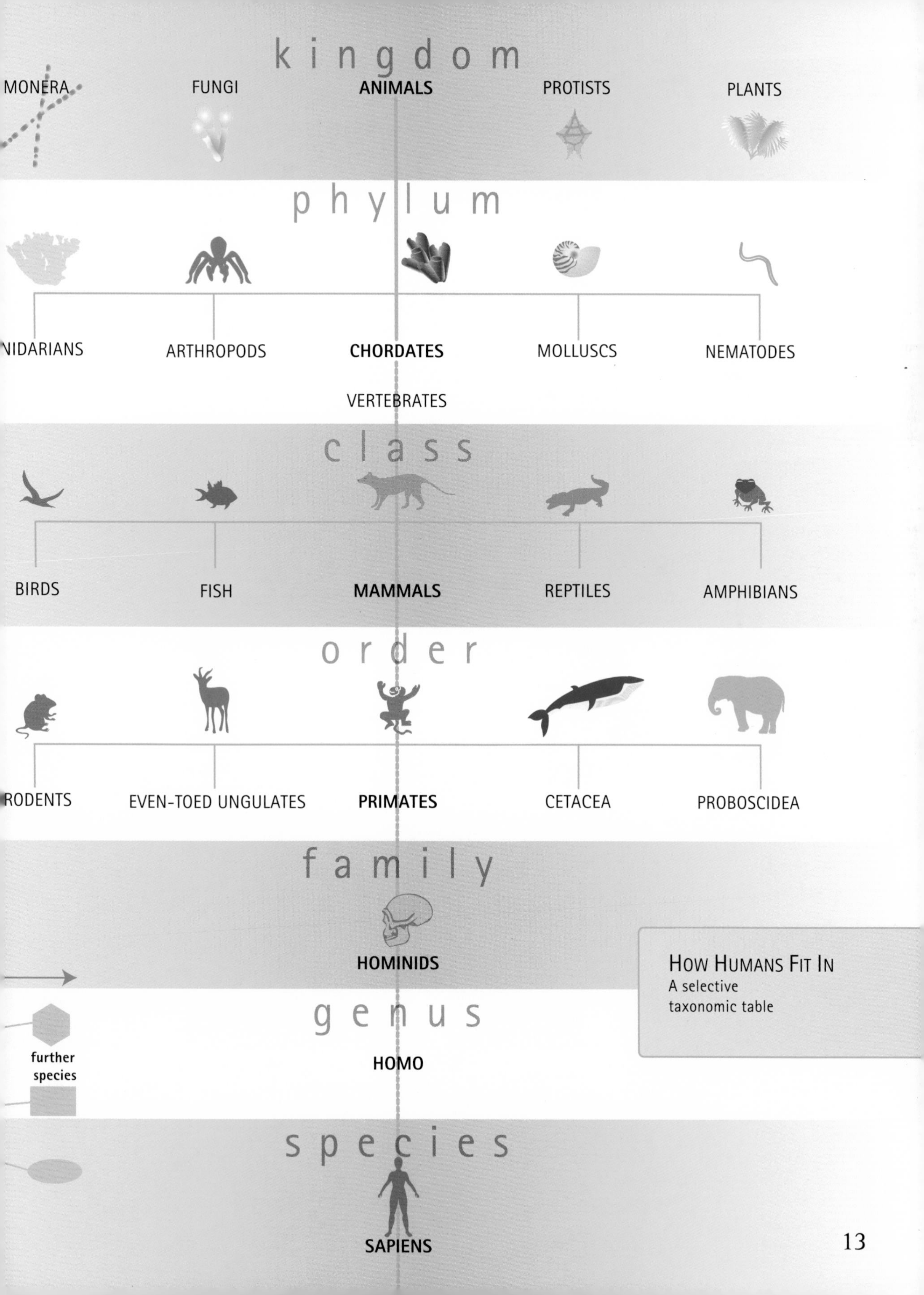

How Humans Fit In
A selective taxonomic table

Mass Extinctions

The "biodiversity" of a place or region is a measure of the number of species present and of their abundance. Each species occupies a "niche" – a specific location, relationship with its physical surroundings and mode of interaction with other species. Organisms of a single species in one place are called a population, and together populations form a community.

Species naturally become extinct as they fail to reproduce, either through extreme conditions or because of displacement by competitors. Even if a species adapts to these threats, it will, by definition, have evolved into a different species. Fossils suggest that this background rate of extinction is punctuated by short periods of mass extinction, defined as a period in which at least 50 percent of all species become extinct.

The precise cause of each mass extinction is difficult to assess. Catastrophes such as meteorite impacts and comet showers may have been responsible for some mass extinctions. Global climate change, fluctuation in the concentration of various gases, and other gradual environmental trends may have caused others.

Interactions between species could even have been a factor, creating instability in complex, finely balanced communities. The loss of certain "keystone" species can be particularly damaging to communities. For example, in areas where otters have recently been hunted nearly to extinction, sea urchins have multiplied, consuming the kelp and so radically altering the habitat. Keystone species may be as small as soil invertebrates or even microbes.

Mass extinctions overwhelmed creatures living both on land and in the sea. On land, animals seem to have suffered more than plants. In the sea, trilobites – a group of marine animals with hard shells – suffered a repeated loss of species.

Mass extinctions have occurred roughly every 26 million years. It is not clear if this pattern has arisen by chance, or whether there is some explanation for it. There have been five particularly widespread mass extinctions. The species that survived, subsequently diversified to occupy the niches vacated by those that vanished. Thus, new forms of life appeared.

Although 1.8 million living species have been

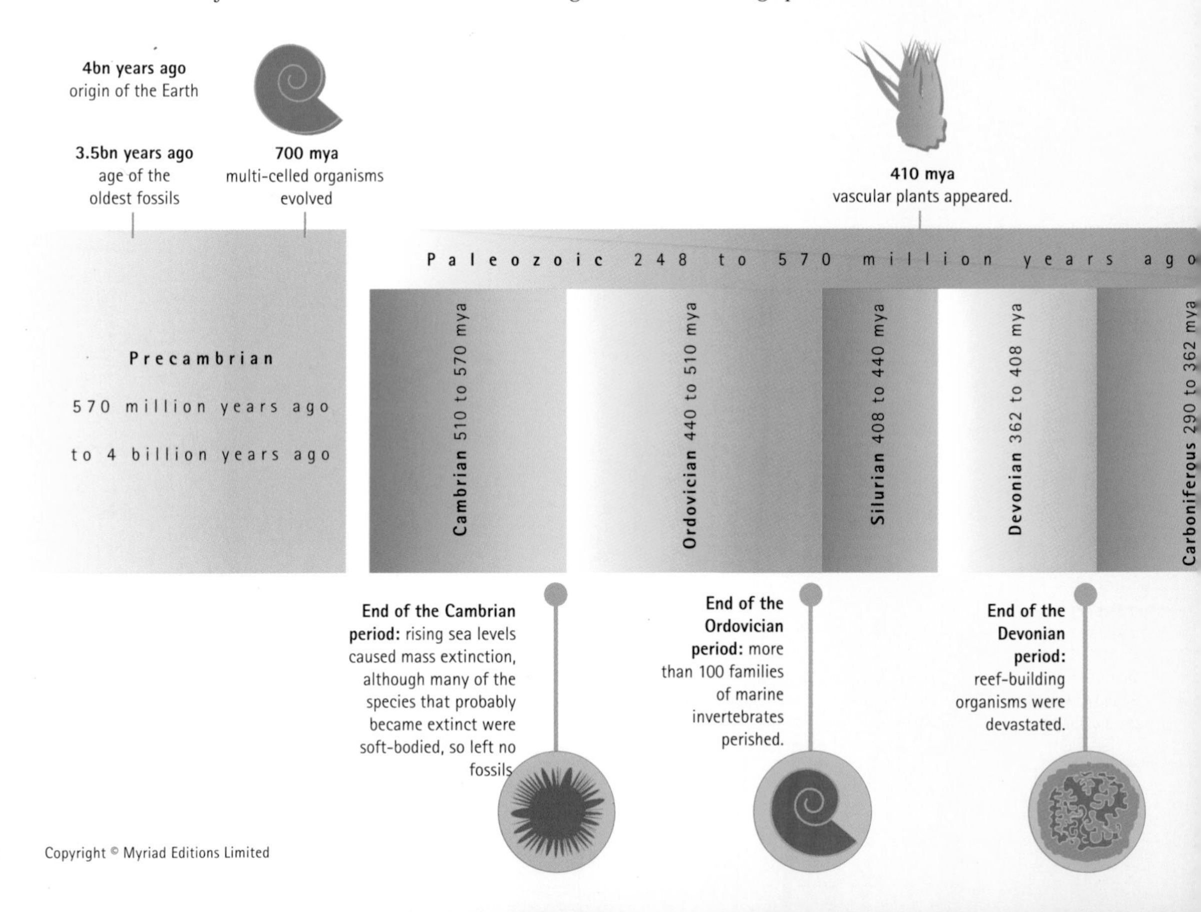

named by scientists, this is but a small fraction of the estimated 10 million to 100 million species thought to be alive on Earth. Most of these are likely to be destroyed by humans before they have even been identified, either as a direct result of human activity such as logging, and pollution, or from the effect on habitat of the rapid climate change resulting from the build-up of greenhouse gases in the upper atmosphere. Many scientists now consider that we are entering what could be the period of the sixth great mass extinction.

Timeline of Mass Extinctions

Main eras and periods

mya millions of years ago

mass extinction

240 mya
dinosaurs, mammals, pterosaurs (flying reptiles), amphibians (including frogs and turtles) appeared.

100,000 years ago
homo sapiens appeared

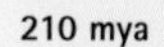

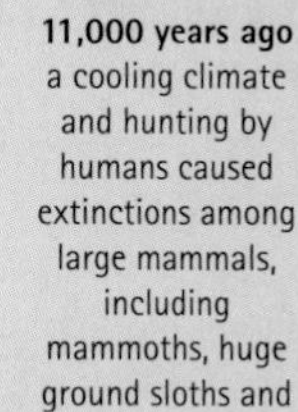

210 mya
towards the end of the Triassic period, climate change caused an extinction of amphibians, marine reptiles and other species, setting the stage for the ascendance of dinosaurs.

190 mya
birds appeared

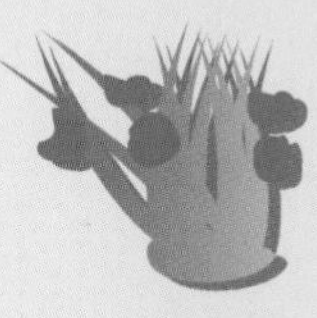

130 mya
flowering plants (angiosperms) appeared

50 mya
primates evolved

11,000 years ago
a cooling climate and hunting by humans caused extinctions among large mammals, including mammoths, huge ground sloths and saber-toothed tigers.

Mesozoic 65 to 248 million years ago

Cenzoic 0 to 65 mya

Permian 248 to 290 mya

Triassic 206 to 248 mya

Jurassic 145 to 206 mya

Cretaceous 65 to 145 mya

Tertiary 2 to 65 mya

End of the Permian period: between 90% and 95% of marine species disappeared, including trilobites.

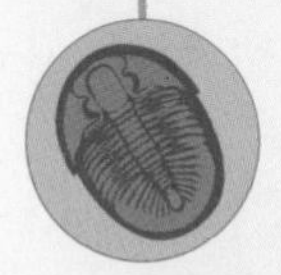

End of the Cretaceous period: 85% of all species, including the dinosaurs, became extinct. The presence of iridium in 66-million-year-old rocks suggests that a meteorite strike shrouded the planet in dust and darkness. Alternatively, the extinction could have been caused by gradual climate change.

Present: In the 20th century the rate of extinction increased to 1% each year – about 10,000 times higher than before human technological society.

HOMININS

Modern humans are the last survivors of the Hominini tribe, whose many branches evolved as a result of geographic isolation, possibly brought about by climate change.

Humans are the only primates who habitually stand and move in an upright position. The oldest known hominins are the genus *Australopithecus*, which evolved from more ape-like creatures over 4 million years ago in Africa and appears not to have spread beyond that continent. Their jaws were ideally suited to feeding on fruit, nuts and berries, and they became extinct around one million years ago probably because their specialized diet became scarce as the climate became drier and savannah replaced woodland.

By the time of its demise, *Australopithecus* was not the only hominin in existence. Two and a half million years ago a larger-brained hominin developed, and evolved into a species (*Homo sapiens*) that was to dominate the planet.

Hominin skulls increased in size as the species evolved, indicating a growing mental capacity. From the size and form of different brain areas, scientists have been able to deduce how intelligence has developed: the frontal lobe indicates abstract processing and a capacity for language whilst the parietal lobe, at the top of the head, is responsible for technological and computational thinking.

It is this use of tools and technology, from clothing and housing to sophisticated farming methods, that has served as a buffer between *Homo sapiens* and the changing environment. Alterations in the population gene pool that lead eventually to new species developing have not been identified in humans for the last 250,000 years.

About two and a half million years ago hominins began fashioning bladed tools by chipping rocks. The same ability for technological and computational thinking now presents opportunities for genetic manipulation and artificial body-parts – a development that may lead to the most rapid, awesome and potentially terrifying period of hominin evolution yet witnessed.

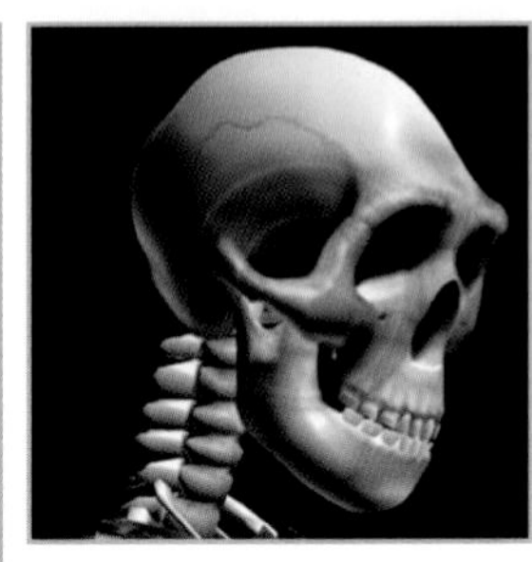

Homo neanderthalensis lived in Southern Europe 100,000 years ago. Up until about 30,000 years ago, they co-existed with modern humans. Their demise may have been due to competition for resources with humans or it is possible they met a more violent end. Alternative theories suggest that Neanderthals were adapted to cold conditions and disappeared as the climate grew warmer. Although some scientists argue that Neanderthals gave rise to modern humans, most agree that they are not our direct ancestors.

Hominins are human-like primates, descended from the genus **Homo**. Bones of *Homo erectus* have been discovered across South and Southeast Asia. They became extinct about 150,000 years ago – before the appearance of *Homo sapiens* around 120,000 years ago.

FOSSIL FINDS IN AFRICA

area where fossils have been found
A Australopithecus
H Homo

H erectus
H erectus
A bahrelghazali
A afarensis
A anamensis
A afarensis
A boisei
H erectus
H habilis
H rudolfensis
A africanus
A robustus
H habilis

Most fossil finds are just fragments, such as a tooth or a piece of skull. **"Lucy"** is special because almost half of her skeleton was recovered and DNA testing showed that she lived 3.2 million years ago.

Lucy was discovered in the Hatar Valley in Ethiopia by Donald Johanson in 1974, and named after the Beatles' song, "Lucy in the Sky with Diamonds" (remade by Elton John that year). Lucy was probably around 25 years old when she died. She would have stood about 3 feet 6 inches (1.10 meters) tall, although other *Australopithecus afarensis* from this area were up to 5 feet 8 inches (1.70 meters) tall.

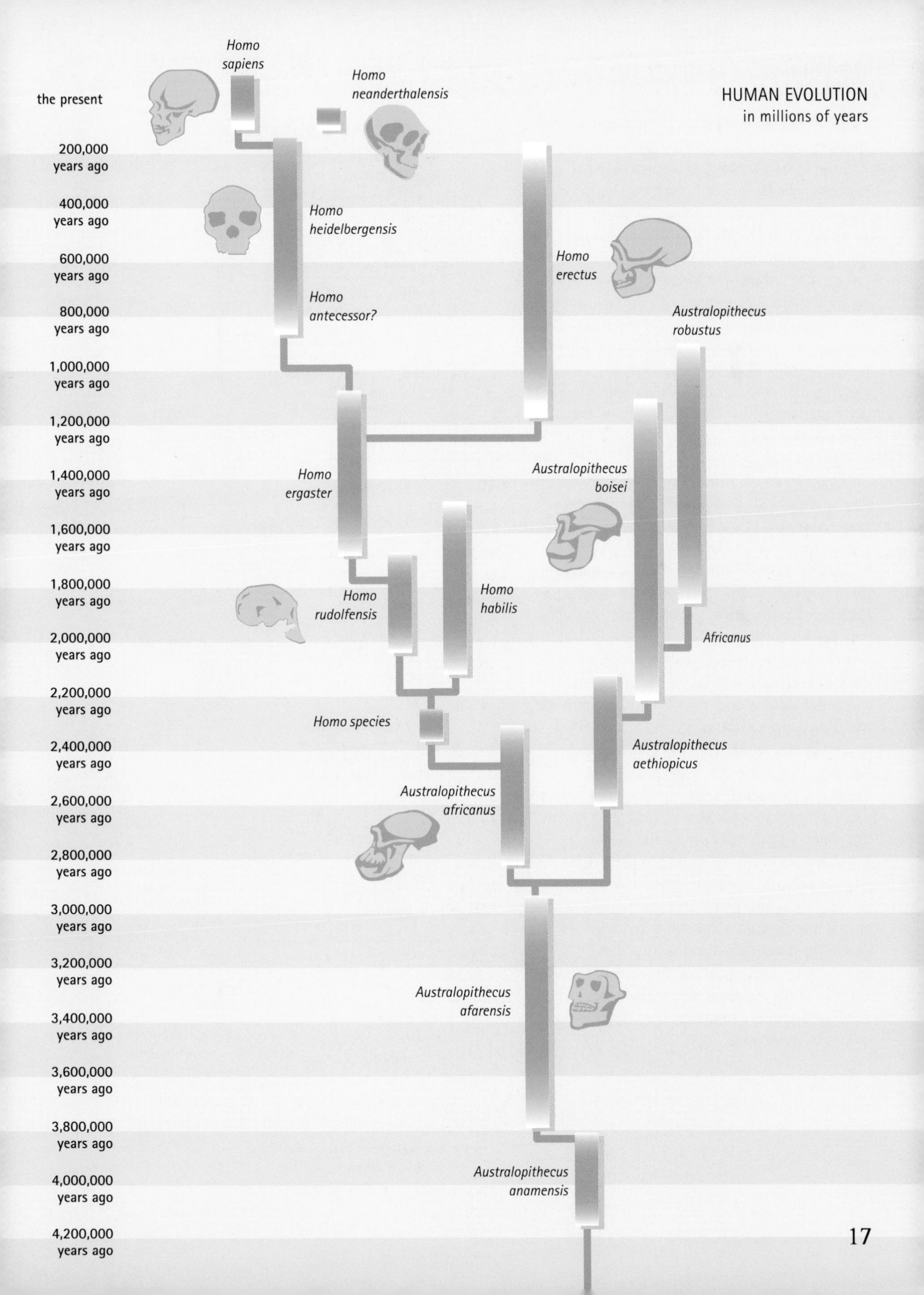
HUMAN EVOLUTION
in millions of years
the present
200,000 years ago
400,000 years ago
600,000 years ago
800,000 years ago
1,000,000 years ago
1,200,000 years ago
1,400,000 years ago
1,600,000 years ago
1,800,000 years ago
2,000,000 years ago
2,200,000 years ago
2,400,000 years ago
2,600,000 years ago
2,800,000 years ago
3,000,000 years ago
3,200,000 years ago
3,400,000 years ago
3,600,000 years ago
3,800,000 years ago
4,000,000 years ago
4,200,000 years ago
Homo sapiens
Homo neanderthalensis
Homo heidelbergensis
Homo antecessor?
Homo erectus
Australopithecus robustus
Homo ergaster
Australopithecus boisei
Homo rudolfensis
Homo habilis
Africanus
Homo species
Australopithecus aethiopicus
Australopithecus africanus
Australopithecus afarensis
Australopithecus anamensis

The total human demand on the planet's natural resources exceeds the Earth's capacity by 25 percent. Since emerging as the dominant species of the *Homo* genus, around 30,000 years ago, *Homo sapiens* has become the most powerful, creative, and ultimately destructive, life-form on the planet. The rate at which humans have multiplied and congregated into vast urban sprawls makes such demands on the environment that in places it is cracking under the strain.

Mining, the burning of fossil fuels and other polluting industrial activities, intensive agriculture involving excessive chemicals, deforestation for timber and agriculture, the creation of road networks that fragment ecosystems, the paving of the Earth's surface, and the over extraction of water from aquifers – all these activities are changing the environment irrevocably. The delicate balance between the species that make up the Earth's rich and varied ecosystems has been disrupted, and thousands of life forms are being pushed towards extinction by the activities of a single species.

By destroying ecosystems, we are sowing the seeds of our own destruction, because it is on these very ecosystems that we depend for our survival. Forests absorb carbon dioxide and stop soil from being washed away, mangroves filter impurities and protect the coastline, oceans help regulate the climate.

Signs of a warming climate are apparent across the world, and the change is occurring so rapidly that many species will have no time to adapt. Rising temperatures are leading to less predictable weather patterns, and more extreme weather events. This is interfering with plant growth and animal breeding, often upsetting mutually dependent relationships and putting many species, not least humans, under increased strain. The way of life – indeed the very survival – of millions of people and hundreds of thousands of species, is threatened.

Climate Change

The effect of global warming on ecosystems and species
2008 or latest available data

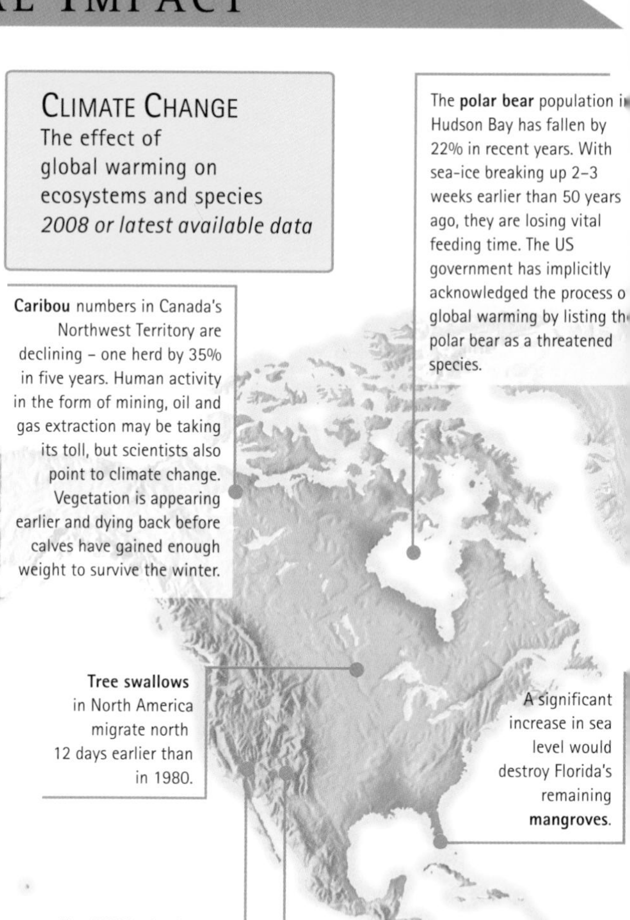

If **global temperatures** increase by 2.5°C or more, up to 30% of species are likely to be at risk of extinction, according to the International Panel on Climate Change.

Arctic sea ice in the summer of 2007 shrank to its smallest extent since satellite records began in 1979 – 23% below the previous record, set in 2005.

ECOLOGICAL FOOTPRINT
The area of productive land or sea needed to resource the lifestyle of one person
2003
hectares

British birds breed on average 9 days earlier than in the mid-20th century. **Frogs** mate up to 7 weeks earlier.

Spring has arrived in Europe earlier by, on average, 3 days every decade since the mid-20th century. This affects not only plant growth, but animal breeding cycles, and bird migratory patterns.

An increase in temperature of 1°C since the 1970s is causing **glaciers** to retreat across the Himalayas. This brings the risk of devastating floods and presents a long-term threat to river flow and water supplies in surrounding countries.

Flooding in Ethiopia in 2006 was followed by an outbreak of cholera.

Flooding in eastern Africa in 2007 created new breeding sites for mosquitoes, increasing levels of malaria, and triggering epidemics of Rift Valley Fever.

A link has been found in Bangladesh between the prevalence of **cholera** and increasingly extreme weather events.

The breeding season of **lemurs** on Madagascar is no longer synchronized with the growing season and the availability of food.

In 1997 **malaria** was detected for the first time as high as 6,900 ft (2,103 m) in Irian Jaya, Indonesia.

Hot weather during **turtle** egg incubation increases the number of female baby turtles. Rising temperatures, coupled with the loss of shade-giving trees due to increasingly violent and frequent storms, has led to a preponderance of female turtles worldwide. No-one knows how this is likely to affect turtle numbers.

The rising sea temperature in the Great Barrier Reef and other **coral reefs** is leading to "bleaching" episodes. A potentially greater threat comes from the reduction in carbonate ions in the oceans caused by increased acidity as a result of the absorption of CO_2.

ECOSYSTEMS

"Try to imagine the Earth without ecosystems...
Each ecosystem represents a solution to a particular
challenge to life, worked out over millennia...
Stripped of its ecosystems, Earth would resemble
the stark, lifeless images beamed back from Mars..."

– *World Resources 2000–2001*

Tropical Forests

The area bounded by the Tropic of Cancer (23.5°N) and the Tropic of Capricorn (23.5°S) is known as "the tropics". Tropical rainforest, mainly comprised of evergreen trees, thrives in areas of high rainfall. Deciduous trees, which shed their leaves in the dry season, tend to grow in areas where there is a seasonal drought. In dry areas that are prone to fire, and where soils are particularly poor, trees grow sparsely, forming "woody savanna".

Tropical forests bind the soil, helping to prevent erosion and retain the few nutrients present. They also absorb carbon dioxide from the atmosphere, exchanging it for oxygen. Tropical forests cover only 6 percent of the Earth's land area but probably contain over half the world's species, most of which have yet to be identified.

Fire, often caused by lightning, is part of the natural cycle of regeneration in tropical forests. Fire destroys the forest canopy, enabling light to reach the ground, and saplings and herbs to grow. In most cases, however, forest fires are caused by humans. In some cases they are deliberately lit in order to clear forests for agriculture. In others, they are started inadvertently, perhaps by discarded cigarettes or sparks from machinery.

Clearing by slashing and burning techniques, and more recently by machines, may completely destroy the remaining major areas of tropical forests. Most of the tropical forests are in poor countries whose priorities are to clear land for agriculture. Attempts by wealthier countries to encourage development often lead to tropical forests being exploited for timber.

Once the forest is gone, the land is prone to erosion and flooding. Nutrients in the soil are rapidly leached and after a few years farming and ranching become untenable. Where the original forest cover is substantial, as in the Amazon, it has a major influence on climate. Loss of trees reduces local rainfall and may even disrupt the global climate. Sustainable exploitation of tropical forests, including selective logging and harvesting of products such as fruit and rubber, can provide a higher long-term income than more destructive practices.

Tropical Forest

Total area of tropical forest
1999 or latest available data
square kilometers

1,350,000 – 3,013,000

500,000 – 890,000

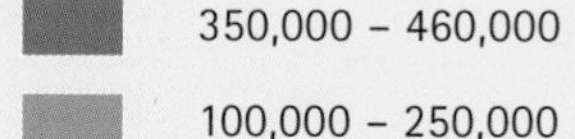

350,000 – 460,000

100,000 – 250,000

10,000 – 99,000

fewer than 10,000

no tropical forest

Protected areas

50,000 square kilometers and/or 40% of total area are protected

The **mahogany tree** is prized worldwide for its dark timber, and about 40% of mahogany logged in Brazil is exported to make furniture. The trees grow slowly, at low density, and the mahogany is already extinct in Honduras and Colombia.

TROPICAL FOREST BY REGION

Protected tropical forest as percentage of tropical forest in the region
1999 or latest available

tropical forest area in thousand hectares

percentage protected

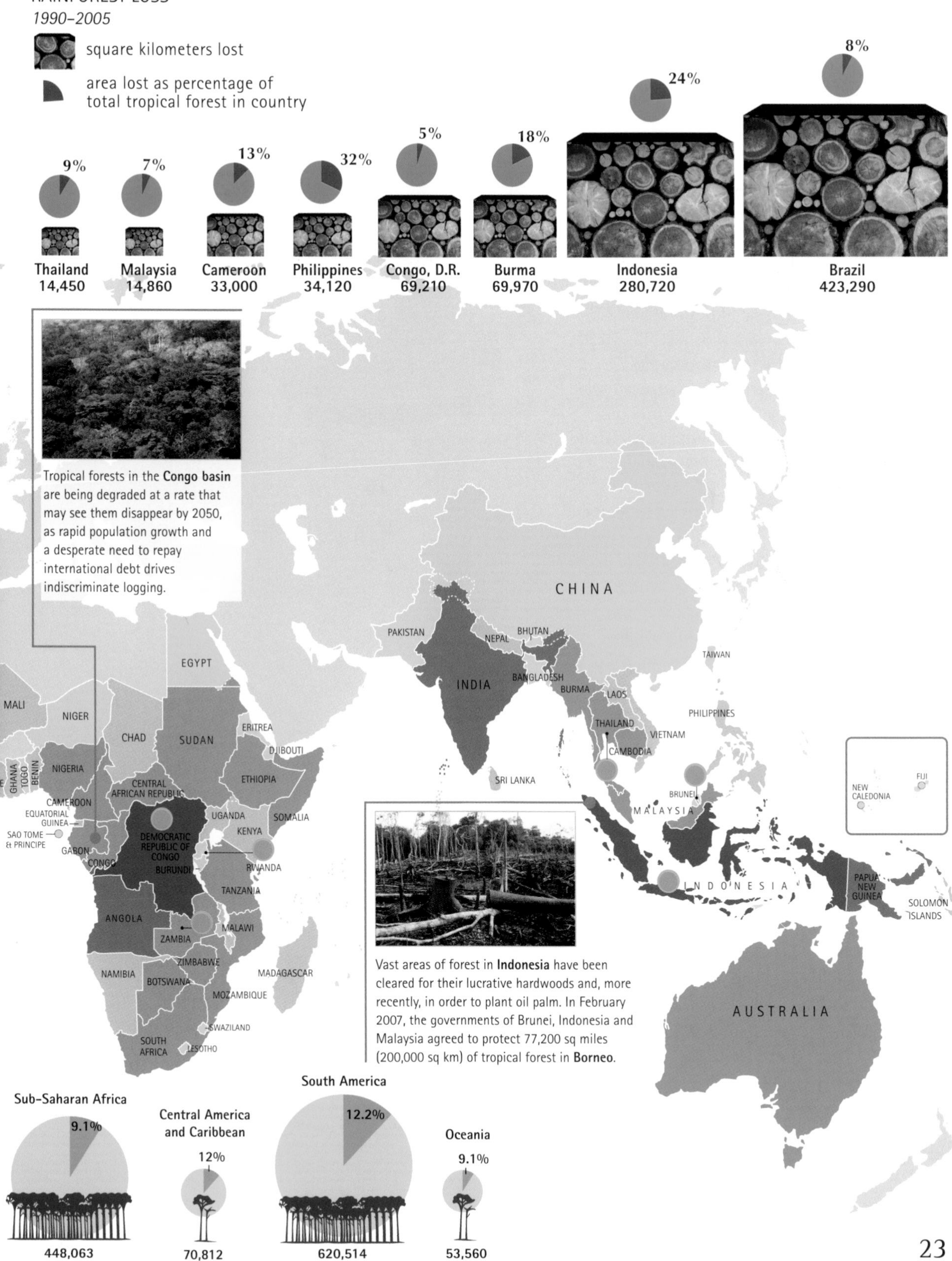

Tropical forests in the **Congo basin** are being degraded at a rate that may see them disappear by 2050, as rapid population growth and a desperate need to repay international debt drives indiscriminate logging.

Vast areas of forest in **Indonesia** have been cleared for their lucrative hardwoods and, more recently, in order to plant oil palm. In February 2007, the governments of Brunei, Indonesia and Malaysia agreed to protect 77,200 sq miles (200,000 sq km) of tropical forest in **Borneo**.

Temperate Forests

Temperate regions lie to the north and south of the tropics (23.5° latitude). Deciduous temperate trees shed their leaves each winter. Nearer the poles "boreal" forests of conifers cover the largest area of any forest type in the world. Conifers also dominate temperate forests at higher altitudes and in places where soils are poor.

Temperate forests have been cleared for agriculture and felled for timber for thousands of years. In the late 20th century this trend slowed as populations in temperate regions stabilized and resources other than timber were used for fuel and building materials. In the 1990s the area of temperate forest cover worldwide rose. However, the ecological quality of many forests has continued to decline as plantations replaced "old growth" forests and the frequency of fires has risen. Fires started by people consume an average of one percent of the existing Mediterranean forest every year.

Inappropriate planting of trees to act as "carbon sinks", a measure to counteract global warming, can actually reduce biodiversity. Most plantations are comprised of a few or even a single tree species such as eucalyptus and pine that may be alien to that habitat. Plantations do, however, make an indirect contribution to conservation by relieving the need to log "old growth" forests. Some plantations are now certified under schemes that demonstrate they are sustainably managed (see pages 96–97).

When forests become fragmented the plants and animals in each forest fragment can be isolated from adjacent forests and local extinction may result. It also becomes more difficult for species to migrate in response to climate change.

Pollution has also damaged temperate forests. The burning of fossil fuels releases sulfur and nitrogen into the atmosphere. Transported by the wind this can fall to earth as dry deposits or combine with water to form "acid rain". Polluting gases are also converted by sunlight into ozone, which interferes with the biological functioning of plants. Forests hundreds of miles from industrial centers are affected, in particular those in the northeast of North America, in East Asia and in northeastern Europe and Scandinavia.

TEMPERATE FOREST BY REGION
Protected temperate forest as percentage of temperate forest in the region *1996*

Russia has the largest area of temperate forest – around 3 million sq miles (8 million sq km) – but less than 1% of it is covered by a certification scheme. Russia has its own controls on logging, but illegal activity is hard to control.

Since 1998, **China** has undertaken a vast reforestation project – and by 2004 more than 177,600 sq miles (460,000 sq km) had been planted – mainly by farmers or other citizens, rather than by the government.

GRASSLANDS

Grassland covers more of the Earth's surface than any other type of terrain. Estimates range from 31 percent to 43 percent, depending on the definition of "grassland", which can include savanna, prairie, scrub, high-altitude plains and Arctic tundra. More than 70 percent of some African countries are covered in grassland, and grassland makes up more than half the terrain in around 40 countries. One of the defining characteristics of grassland is that its vegetation is prevented from turning into forest by fire, grazing, lack of water or by freezing temperatures.

Grasslands support a wide range of wild animals, in particular birds. As well as providing year-round habitat for endemic bird species, grasslands also provide temporary refuge and breeding sites for migrating birds. They are therefore of vital importance, and their degradation in North America has resulted in a declining bird population since the mid-1960s.

Grasslands also provide grazing for domestic animals. In some areas of the world they have supported nomadic herds of sheep, goats and cattle for thousands of years. With growing human populations rearing ever more livestock, however, grasslands are at risk from over-grazing, which leads to soil erosion.

One of the major threats to grassland ecosystems is fragmentation by development, including roads. Breaking grassland up into small patches reduces its capacity to maintain biological diversity – it becomes degraded. This is what is happening in the Great Plains of the USA, where criss-crossing roads have fragmented 70 percent of the area into patches smaller than 386 square miles (1,000 square kilometres).

Much of the watershed areas of many of the world's largest rivers are comprised of grassland. Here it performs the vital function of absorbing rainfall into underground aquifers that in turn feed into the river systems. Grasslands also act as "carbon sinks" by absorbing carbon dioxide from the atmosphere. They hold an estimated 33 percent of carbon stored in terrestrial ecosystems, most of which is to be found in the soil.

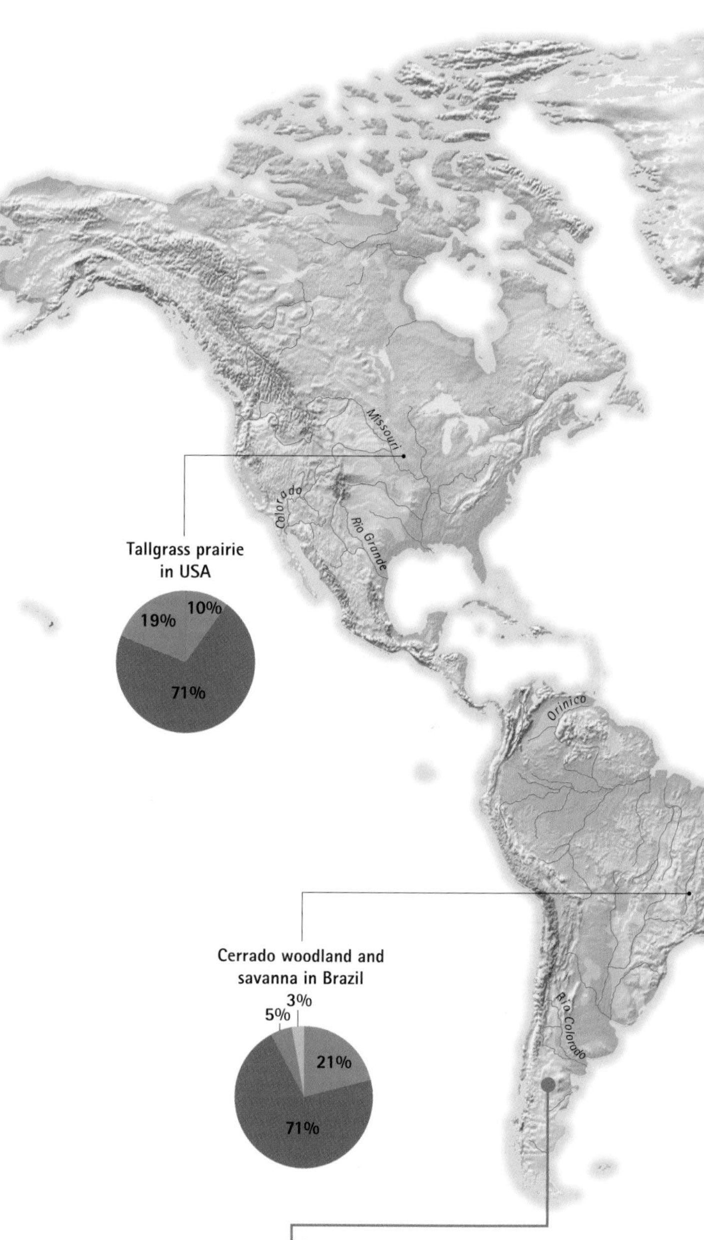

The dry **Patagonian steppe** in Argentina hosts abundant wildlife, including the endemic wild llama, the "guanaco". Human settlement is limited to "estancias" (ranches) and a few small towns. Its aridity leaves the Patagonian steppe vulnerable to overgrazing by sheep and goats, which are turning some areas into desert. Pumas are hunted, often illegally, because they prey on livestock.

The grassland of **Asia's high steppes** supports around 30 million livestock, many of which are grazed on a nomadic system. However, over the past 50 years Russian and Chinese herders have been encouraged to adopt more sedentary grazing methods mixed with arable farming. As a result, the fragile ecology of their grassland has been unbalanced, and around 75% has been degraded. Although the grassland in Mongolia is in better condition it is also now threatened by an increase in livestock grazed as a result of privatization since 1990.

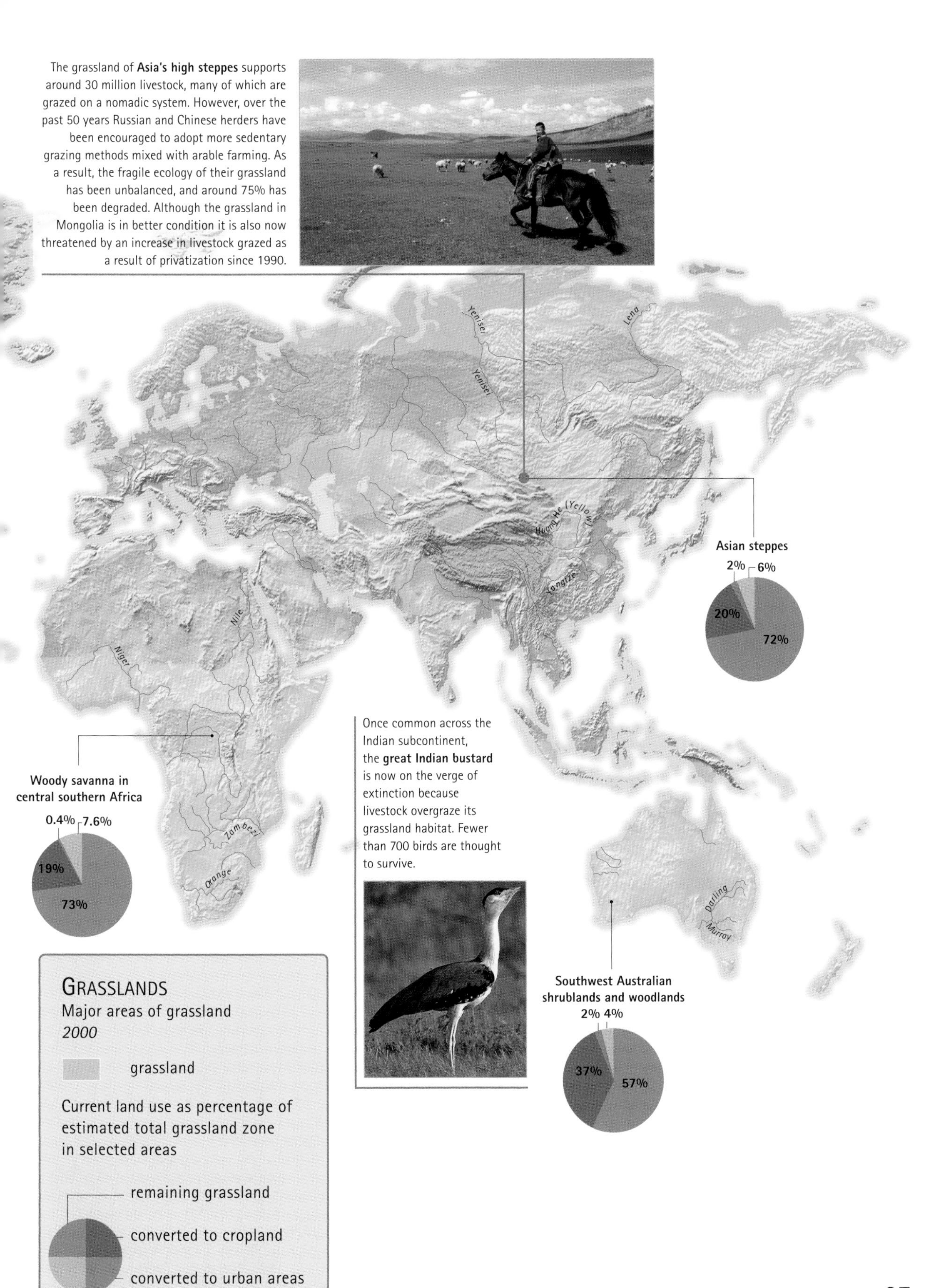

Once common across the Indian subcontinent, the **great Indian bustard** is now on the verge of extinction because livestock overgraze its grassland habitat. Fewer than 700 birds are thought to survive.

GRASSLANDS

Major areas of grassland
2000

grassland

Current land use as percentage of estimated total grassland zone in selected areas

remaining grassland
converted to cropland
converted to urban areas
other

WETLANDS

Wetlands are among the world's richest and most productive ecosystems. They include swamps, marshes, mangroves, lakes and rivers and cover over 2 million square miles (5.7 million sq km).

Wetlands occur on poorly drained land, where organic matter decays very slowly and peat accumulates, creating bogs. At northern latitudes mosses such as *Sphagnum* dominate the flora. Where the soil has more nutrients, both inland and at the mouths of great rivers such as the Mississippi and the Nile, grass grows to form marshes. Mangroves, adapted to grow in muddy tidal water that alternates between salty seawater and fresh, occur around deltas and elsewhere. They are mainly found in the tropics, but where ocean currents are favorable, occur also on sub-tropical coasts.

Salt marshes support a large number of invertebrates, providing food for diverse species of birds. Mangrove forests support lichens, orchids, and bacteria, and provide nesting sites for birds, and vital nursery and feeding sites for fish, crustaceans and other shellfish.

Agriculture is the principal cause of wetland destruction, but the damming of rivers can also disrupt these delicate ecosystems. Many wetlands have also been drained in an attempt to destroy the breeding sites of mosquitoes, and thereby control malaria. Wetlands have already been lost in Europe and North America and losses are now high in Asia and Africa. Inter-tidal salt marshes have been "reclaimed" and mangroves destroyed for the development of ports, marinas, housing and commercial fisheries. But salt marshes and mangroves provide vital ecological functions, helping to stabilize estuary banks and provide a barrier against the sea. Their destruction causes erosion and land subsidence and permits salt water to penetrate coastal soils and threaten fresh water supplies.

The Convention on Wetlands was signed in Ramsar, Iran in 1971. By mid-2008, 1,752 sites were included in the Ramsar List of Wetlands of International Importance, representing over 618,000 square miles (1.6 million sq km) of wetlands, nearly twice that in 2001. Signatories to the Convention undertake to practise "wise use" of those wetlands, to sustain their biodiversity.

FLORIDA MANGROVES

mangroves

boundary of National Park

Tampa
Florida
Lake Worth
Miami
Everglades National Park World Heritage Site
Biscayne Bay National Park
Florida Keys

Florida has an estimated 765 square miles (2,000 sq km) of mangrove forests, comprising three different species: the red, black and white mangrove. During the 20th century large swathes were destroyed as the area was developed, including 44% in Tampa Bay, and 87% of those around Lake Worth. Even the conservation measures adopted in the Everglades National Park are not enough to protect it from water pollution. Florida's wading birds, which depend on mangroves for their nesting areas, have declined to around 10% of their original level.

CANADA
USA
MEXICO
BAHAMAS
CUBA
DOMINICAN REP.
BELIZE
JAMAICA
GUATEMALA
HONDURAS
EL SALVADOR
NICARAGUA
ANTIGUA & BARBUDA
DOMINICA
ST LUCIA
BARBADOS
COSTA RICA
PANAMA
VENEZUELA
TRINIDAD & TOBAGO
SURINAME
COLOMBIA
ECUADOR
PERU
BRAZIL
BOLIVIA
PARAGUAY
CHILE
ARGENTINA
URUGUAY

The financial value of the services mangroves provide in terms of timber and shore protection is hard to assess, but range from \$39,000–\$130,000 per sq mile (\$15,000–\$50,000 per sq km) each year, or as much as \$3 million (\$1m) a year in popular tourist areas.

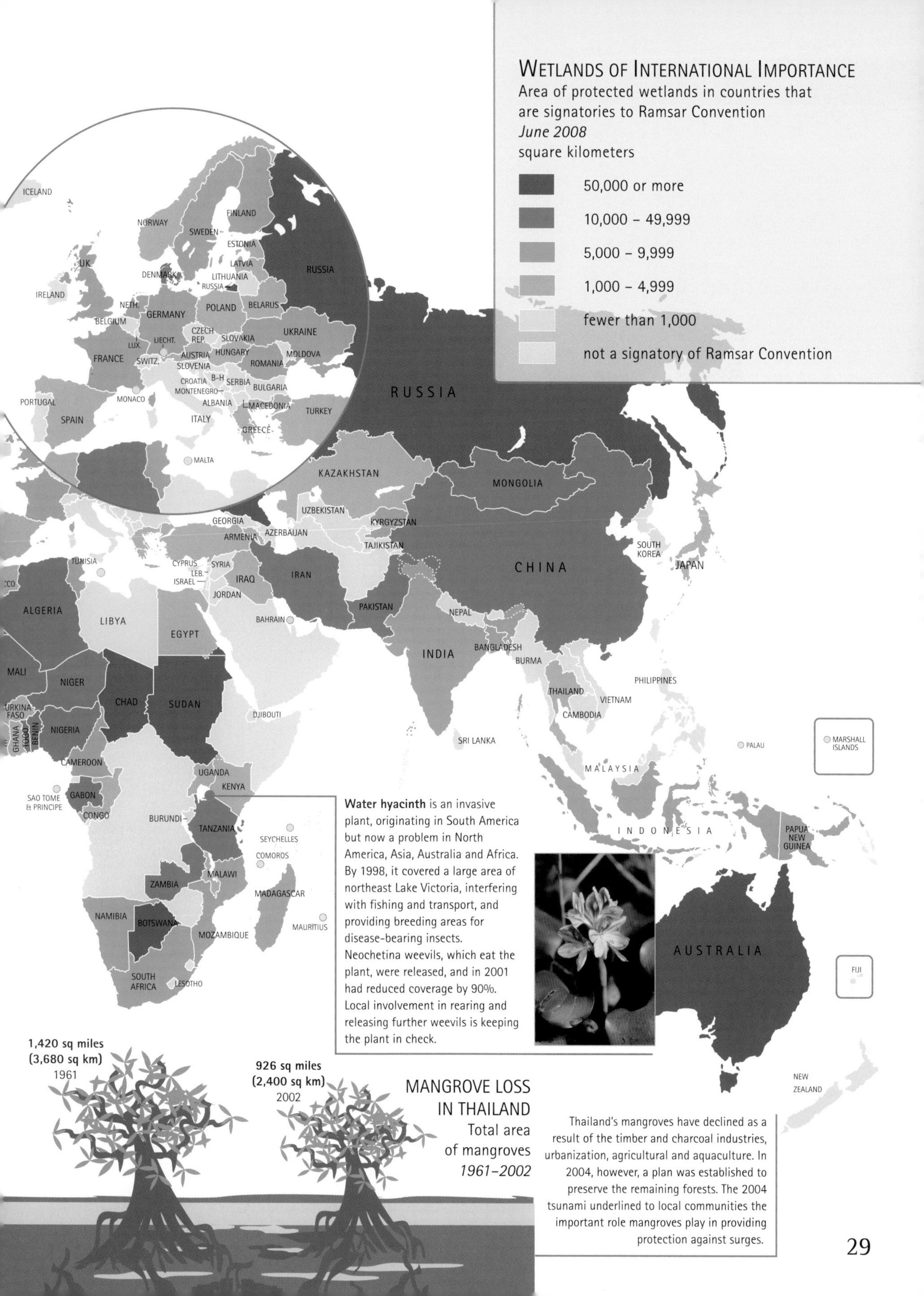
WETLANDS OF INTERNATIONAL IMPORTANCE
Area of protected wetlands in countries that are signatories to Ramsar Convention
June 2008
square kilometers
50,000 or more
10,000 – 49,999
5,000 – 9,999
1,000 – 4,999
fewer than 1,000
not a signatory of Ramsar Convention
ICELAND
NORWAY
SWEDEN
FINLAND
ESTONIA
LATVIA
LITHUANIA
RUSSIA
DENMARK
UK
IRELAND
NETH.
BELGIUM
GERMANY
POLAND
BELARUS
CZECH REP.
SLOVAKIA
UKRAINE
LUX.
LIECHT.
AUSTRIA
HUNGARY
MOLDOVA
FRANCE
SWITZ.
SLOVENIA
ROMANIA
CROATIA
B-H
SERBIA
BULGARIA
MONTENEGRO
PORTUGAL
MONACO
ALBANIA
MACEDONIA
TURKEY
SPAIN
ITALY
GREECE
MALTA
RUSSIA
KAZAKHSTAN
MONGOLIA
UZBEKISTAN
GEORGIA
KYRGYZSTAN
ARMENIA
AZERBAIJAN
TAJIKISTAN
SOUTH KOREA
JAPAN
CHINA
TUNISIA
CYPRUS
SYRIA
LEB.
ISRAEL
IRAQ
IRAN
JORDAN
ALGERIA
LIBYA
BAHRAIN
PAKISTAN
NEPAL
EGYPT
INDIA
BANGLADESH
BURMA
MALI
NIGER
PHILIPPINES
THAILAND
VIETNAM
CHAD
SUDAN
CAMBODIA
DJIBOUTI
GHANA
TOGO
BENIN
NIGERIA
SRI LANKA
PALAU
MARSHALL ISLANDS
CAMEROON
MALAYSIA
UGANDA
KENYA
SAO TOME & PRINCIPE
GABON
CONGO
BURUNDI
TANZANIA
SEYCHELLES
INDONESIA
PAPUA NEW GUINEA
COMOROS
MALAWI
ZAMBIA
MADAGASCAR
NAMIBIA
BOTSWANA
MAURITIUS
MOZAMBIQUE
AUSTRALIA
SOUTH AFRICA
LESOTHO
FIJI
NEW ZEALAND
Water hyacinth is an invasive plant, originating in South America but now a problem in North America, Asia, Australia and Africa. By 1998, it covered a large area of northeast Lake Victoria, interfering with fishing and transport, and providing breeding areas for disease-bearing insects. Neochetina weevils, which eat the plant, were released, and in 2001 had reduced coverage by 90%. Local involvement in rearing and releasing further weevils is keeping the plant in check.
1,420 sq miles (3,680 sq km) 1961
926 sq miles (2,400 sq km) 2002
MANGROVE LOSS IN THAILAND
Total area of mangroves
1961–2002
Thailand's mangroves have declined as a result of the timber and charcoal industries, urbanization, agricultural and aquaculture. In 2004, however, a plan was established to preserve the remaining forests. The 2004 tsunami underlined to local communities the important role mangroves play in providing protection against surges.

SHALLOW CORAL REEFS

Coral reefs are found in the shallow coastal waters of over 100 countries. They grow profusely in warm, well-circulating calm waters and cover an estimated 231,600 square miles (600,000 sq km) worldwide. Home to a vast number of different species, coral reefs are second only to rainforests in species richness.

Coral consists of thousands of invertebrate marine animals – known as "polyps" – with a hollow, cylindrical structure and a skeleton containing calcium carbonate. The lower end of the coral is attached to a rock or another polyp. At the free end is a mouth, surrounded by tentacles that can be extended to paralyze prey. While corals living in deep water (see pages 32–33) rely on this method, corals in shallow water obtain most of their food from photosynthesis by algae living inside them.

Sick coral provides an early warning that entire ecosystems are in danger. Since the 1980s, dozens of new infections, including white band and yellow pox, have attacked corals. Few of these ailments have a known cause, but human development of coastal zones is a likely factor.

Increases in sea temperature and level can cause the coral polyps to shed the algae on which they depend, causing them to lose their colour and die ("bleaching"). Although some reefs do recover over time, if climate change makes such episodes more frequent, this will no longer be possible.

The rising concentration of carbon dioxide in the atmosphere means that more is being absorbed by the world's oceans. This is causing the water to become more acidic, which will eventually make it impossible for organisms, including corals, to form shells. Scientists are now warning that this might occur in surface waters as early as 2050.

Illegal fishing methods, such as the use of cyanide and explosives to stun fish, damages coral. Over-fishing also affects the ecological balance of a reef, which can become overgrown with algae if grazing fish are removed.

The loss of coral reefs is likely to reduce the fish catch of many tropical developing countries, around 25 percent of which comes from reef environments. Where reefs may have acted as barriers against erosion, their destruction may also allow the sea to encroach on coastal regions.

Nearly a third of 704 species of living coral studied are under threat of extinction, according to a 2008 IUCN report, which drew on the conclusions of scientists from 14 nations.

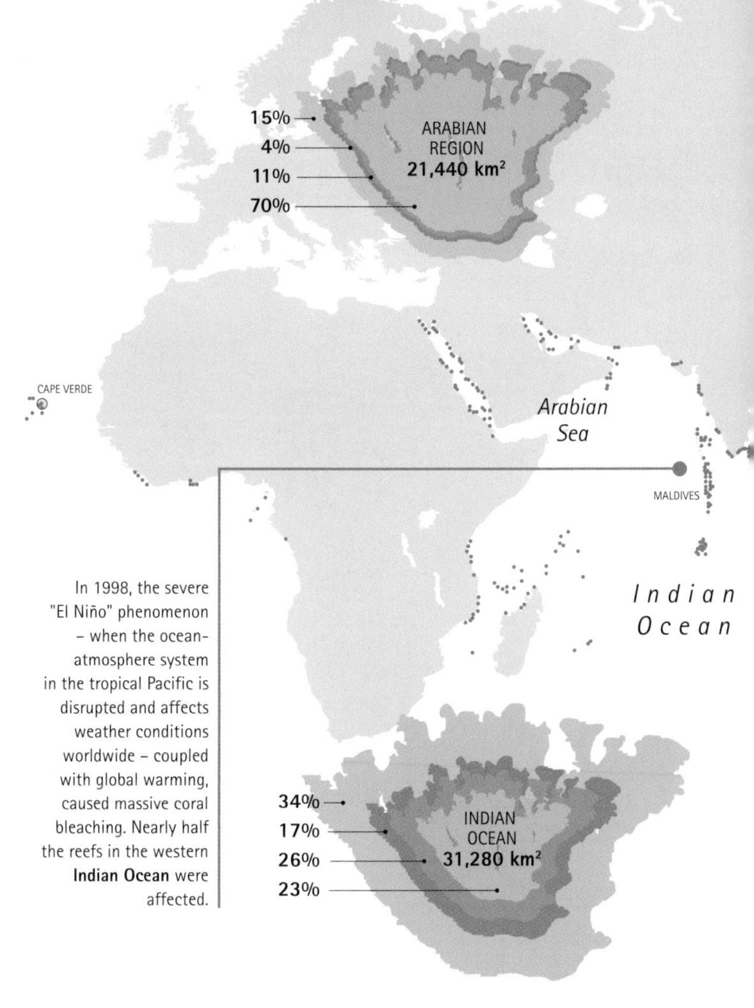

In 1998, the severe "El Niño" phenomenon – when the ocean-atmosphere system in the tropical Pacific is disrupted and affects weather conditions worldwide – coupled with global warming, caused massive coral bleaching. Nearly half the reefs in the western **Indian Ocean** were affected.

Over 350 protected areas include coral reefs, but these are often in countries without adequate resources to enforce the necessary controls. Tourism can both harm coral reefs, and supply the incentive and finances to protect reefs, but it has to be carefully managed.

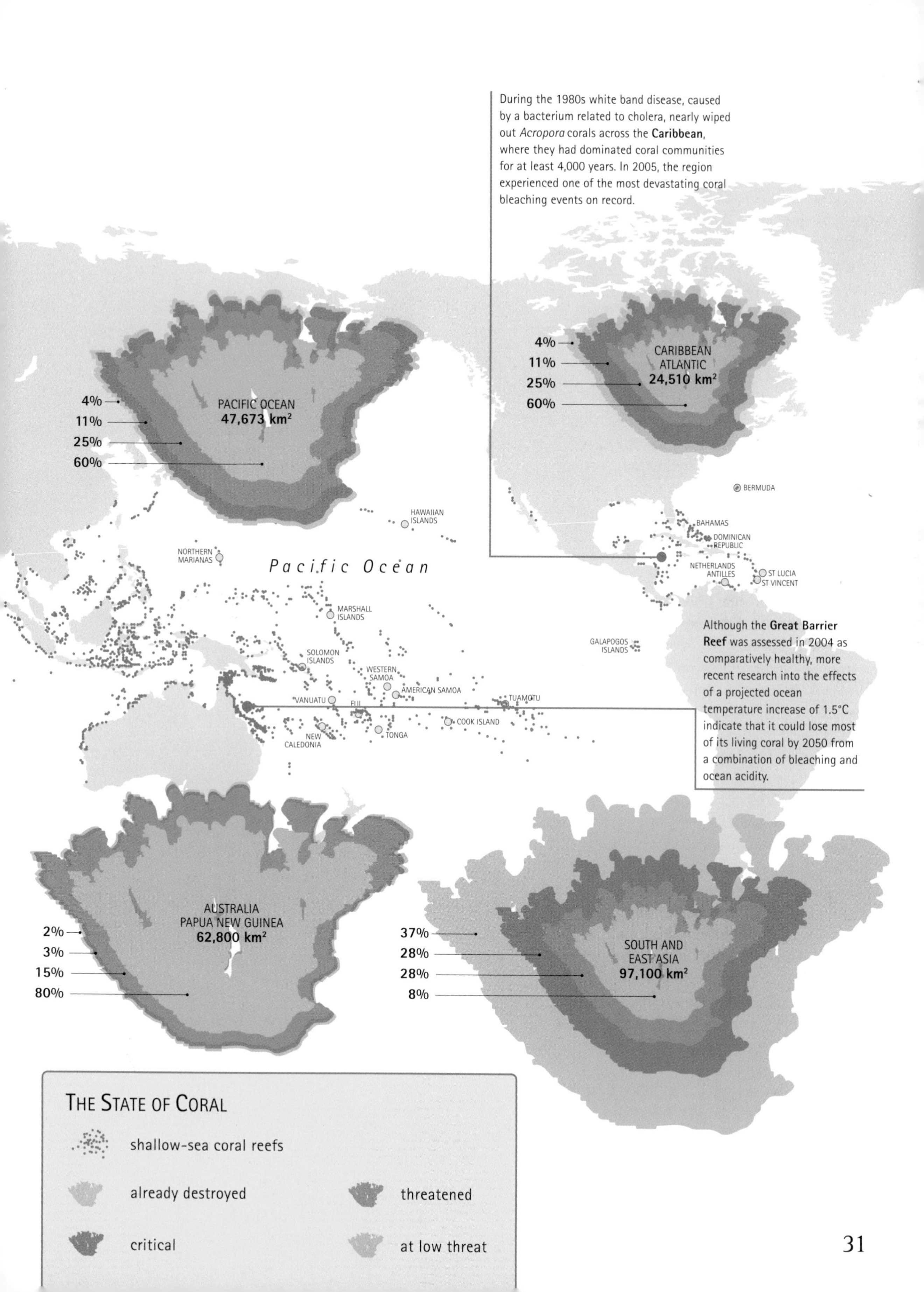
During the 1980s white band disease, caused by a bacterium related to cholera, nearly wiped out *Acropora* corals across the **Caribbean**, where they had dominated coral communities for at least 4,000 years. In 2005, the region experienced one of the most devastating coral bleaching events on record.
4%
11%
25%
60%
PACIFIC OCEAN
47,673 km²
4%
11%
25%
60%
CARIBBEAN
ATLANTIC
24,510 km²
BERMUDA
HAWAIIAN ISLANDS
BAHAMAS
DOMINICAN REPUBLIC
NORTHERN MARIANAS
Pacific Ocean
NETHERLANDS ANTILLES
ST LUCIA
ST VINCENT
MARSHALL ISLANDS
Although the **Great Barrier Reef** was assessed in 2004 as comparatively healthy, more recent research into the effects of a projected ocean temperature increase of 1.5°C indicate that it could lose most of its living coral by 2050 from a combination of bleaching and ocean acidity.
GALAPOGOS ISLANDS
SOLOMON ISLANDS
WESTERN SAMOA
AMERICAN SAMOA
VANUATU
FIJI
TUAMOTU
COOK ISLAND
NEW CALEDONIA
TONGA
AUSTRALIA
PAPUA NEW GUINEA
62,800 km²
2%
3%
15%
80%
37%
28%
28%
8%
SOUTH AND
EAST ASIA
97,100 km²
THE STATE OF CORAL
shallow-sea coral reefs
already destroyed
threatened
critical
at low threat

There could be up to 5 million species inhabiting the dark recesses of the sea, in ecosystems that are being destroyed even before they have been discovered. This rich and varied landscape has, until recently, been largely unstudied. Only since the development of submersibles able to withstand immense pressure have some of the secrets of the ocean depths been revealed.

Plankton and the carcasses of larger animals, drawn from the surface by gravity and currents, provide one source of energy in these completely dark waters, and cold-water coral on the seabed survives at depths of over 19,000 feet (6,000 meters), using tree-like branching structures to capture its food. Volcanoes thrust from the thin crust under the oceans, forming "seamounts", and hydrothermal vents in the ocean floor belch out hydrogen sulphide, which microbes harness through chemosynthesis to provide another source of energy.

It is around seamounts that the most abundant deep-sea life is found. But even as these features are being identified and studied, they are being targeted by deep-sea bottom trawlers, which drag nets across the sea-bed, completely destroying any coral in its path. Such fishing practices contribute only around 1 percent of the world's fish production but are wreaking immeasurable environmental damage.

Deep-sea species tend to be long-lived and slow to reproduce, and deep-sea trawling is causing their stocks to decline. Since trawling for orange roughy started in the late 1970s, stocks of this species are thought to have fallen by more than 70 percent in more than half the areas studied. In addition to the fish targeted by the trawlers, the nets bring up many other species, such as rare deep-sea sharks. The UN General Assembly called, in 2006, for urgent action to protect the fragile ecosystems on which these species depend. A meeting of 40 countries, organized by the UN Food and Agriculture Organization early in 2008, failed to draw up guidelines.

Localized damage to, and pollution of, the seabed is caused by drilling for oil and gas. As fields in shallower waters are nearing depletion, and technology is developed to drill at ever-greater ocean depths, companies are prospecting in deeper waters. Another potential source of energy, lying on the seabed itself, are nodules of frozen methane, although the technology for capturing the gas from these methane hydrates is still in its infancy. Interest has also been shown in sources of metals and minerals to be found at great depths on the ocean floor, with escalating prices for raw materials making the mining of these a more viable commercial proposition.

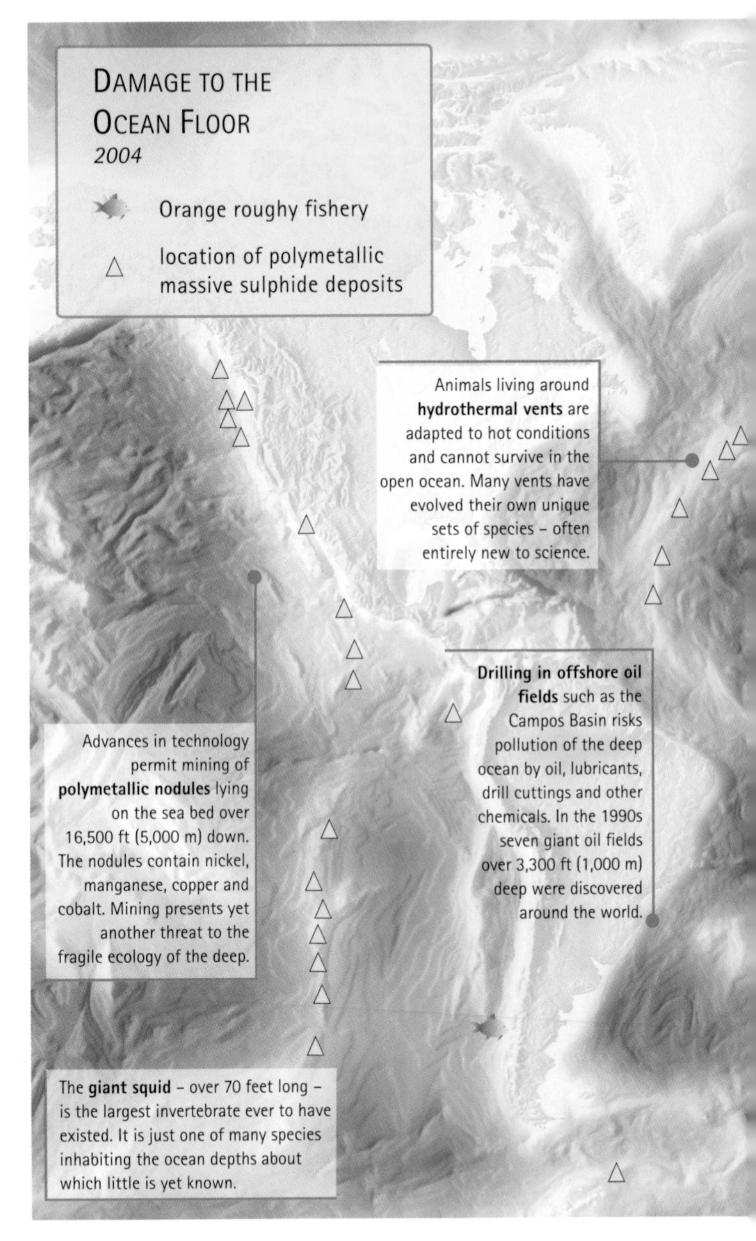

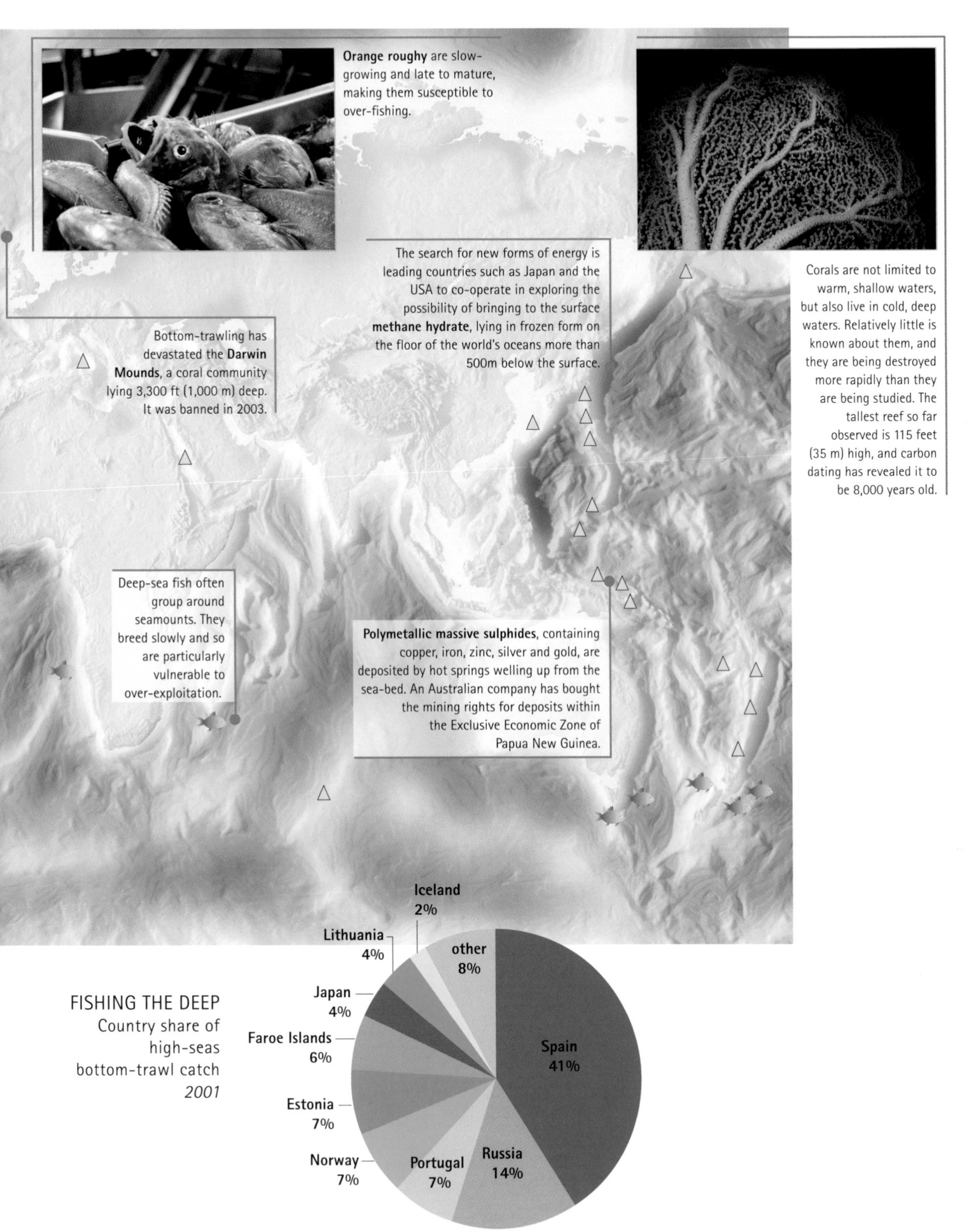
Orange roughy are slow-growing and late to mature, making them susceptible to over-fishing.
Corals are not limited to warm, shallow waters, but also live in cold, deep waters. Relatively little is known about them, and they are being destroyed more rapidly than they are being studied. The tallest reef so far observed is 115 feet (35 m) high, and carbon dating has revealed it to be 8,000 years old.
The search for new forms of energy is leading countries such as Japan and the USA to co-operate in exploring the possibility of bringing to the surface **methane hydrate**, lying in frozen form on the floor of the world's oceans more than 500m below the surface.
Bottom-trawling has devastated the **Darwin Mounds**, a coral community lying 3,300 ft (1,000 m) deep. It was banned in 2003.
Deep-sea fish often group around seamounts. They breed slowly and so are particularly vulnerable to over-exploitation.
Polymetallic massive sulphides, containing copper, iron, zinc, silver and gold, are deposited by hot springs welling up from the sea-bed. An Australian company has bought the mining rights for deposits within the Exclusive Economic Zone of Papua New Guinea.
FISHING THE DEEP
Country share of high-seas bottom-trawl catch
2001
Spain 41%
Russia 14%
Portugal 7%
Norway 7%
Estonia 7%
Faroe Islands 6%
Japan 4%
Lithuania 4%
Iceland 2%
other 8%
Total catch: 205,024 tonnes

Fragile Regions

"There are no passengers on spaceship earth.
We are all crew."

– Marshall McLuhan

The Arctic

The Arctic presents some of the harshest conditions on the planet but is home to unique and fragile communities. Plants that are dormant for most of the year blossom during the brief Arctic summer. Vast numbers of birds, including over a hundred species of waterfowl and waders, breed and then migrate all over the world. Animals, such as polar bears, arctic foxes and seals, breed and then remain to over-winter.

Polar bears used to be widely hunted for their skins and meat. By 1970 their numbers had fallen below 10,000. In 1973, Canada, Denmark (which governs Greenland), Norway, the USA, and what was the USSR signed the International Agreement on Conservation of Polar Bears and Their Habitat. The treaty protects the polar bears' feeding and breeding grounds and their migration routes. It also bans the capture of polar bears, except by scientists working to preserve the species, and by the Inuit, who are allowed to hunt only a certain number each year, and are banned from doing so when bears are pregnant or with their cubs.

The countries have established reserves where polar bears are completely protected. The international agreement also states that all five nations must ban polar-bear hunting from aircraft and large motorized boats, conduct and co-ordinate management and research efforts, and exchange research results and data. Since 1973 the polar bear population has risen again to between 20,000 and 40,000.

Hunting is not the only threat to Arctic wildlife. In recent years, warmer Atlantic Ocean water has penetrated the Arctic Ocean basin. Arctic ice cover and salinity have declined as the ice-cap has melted. Permafrost soils in Alaska, Canada and Russia are thawing. The area of Arctic ice is shrinking as a result of climate change, causing reductions in ice algae, which live beneath the ice and form the base of the Arctic food chain. This will affect fish, seals, whales and polar bears. Polar bears are already suffering from the loss of their hunting grounds on ice-shelves. Global warming will also cause forests to move north, replacing the Arctic tundra, affecting birds, such as the endangered red-breasted goose (see right) that breed in the Russian tundra.

Persistent organic pollutants (POPs) are produced by industries all over the world. They are wafted to the Arctic by wind and sea. Arctic wildlife is exposed to DDT (a pesticide), PCBs (from electronic equipment), and dioxins (from plastics). POPs are not excreted by animals, but accumulate in their fatty tissue. Those highest up the food chain, such as seals and polar bears, are most severely exposed. These chemicals can affect fertility and damage the immune system. Heavy metals, including mercury, arsenic, and lead are also carried to the Arctic.

One of the biggest threats to wildlife comes from drilling for oil and gas, which involves an extensive infrastructure of pipelines, roads, and harbors, all of which can disrupt animal migration routes, and fragment ecologically rich areas. They also bring the risk of oil spillage, as does the increase in shipping in the region that is expected as a result of the shrinking ice-fields. Russia's claim to 463,000 square miles (1.2 million sq km) of land beneath the ice-cap on the grounds that it is an extension of its own continental shelf, if successful, would increase the possibility of under-sea mining in the region.

Collaboration between polar nations to address these wider problems is co-ordinated by the Arctic Council. The deeper causes of many of the problems facing the Arctic, such as global warming and pollution, lie much further to the south, in the industrialized nations, and can only be tackled there.

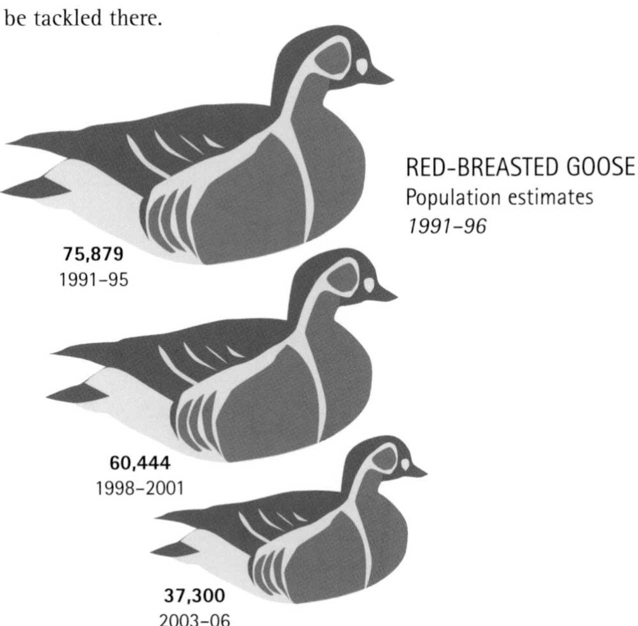

RED-BREASTED GOOSE
Population estimates
1991–96

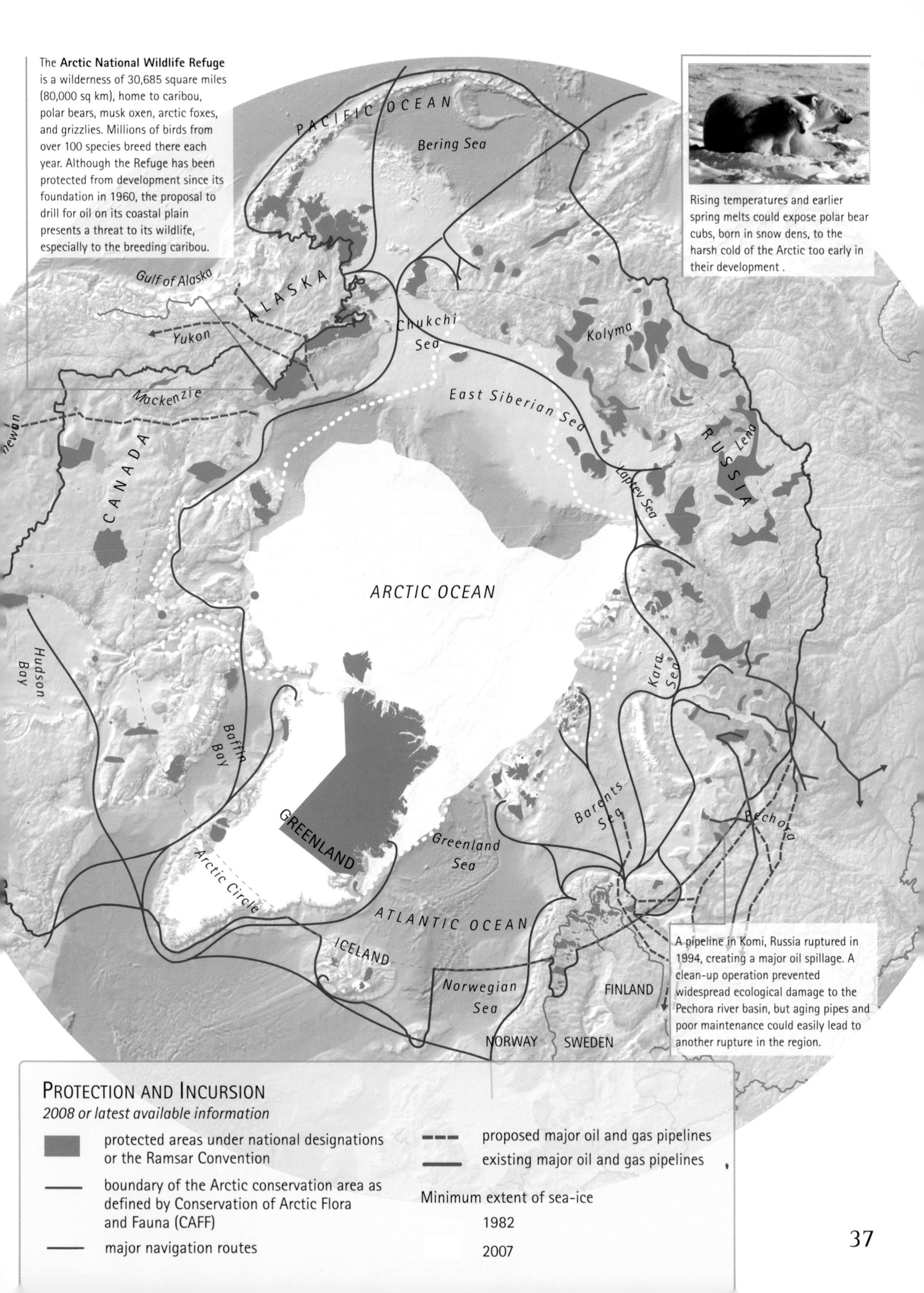

The **Arctic National Wildlife Refuge** is a wilderness of 30,685 square miles (80,000 sq km), home to caribou, polar bears, musk oxen, arctic foxes, and grizzlies. Millions of birds from over 100 species breed there each year. Although the Refuge has been protected from development since its foundation in 1960, the proposal to drill for oil on its coastal plain presents a threat to its wildlife, especially to the breeding caribou.

Rising temperatures and earlier spring melts could expose polar bear cubs, born in snow dens, to the harsh cold of the Arctic too early in their development .

A pipeline in Komi, Russia ruptured in 1994, creating a major oil spillage. A clean-up operation prevented widespread ecological damage to the Pechora river basin, but aging pipes and poor maintenance could easily lead to another rupture in the region.

The Antarctic

The Antarctic is a continent larger than Europe, measuring 5.4 million square miles (14 million sq km). Temperatures rarely rise above freezing, and most of the land is covered in ice. Glaciers flow from the interior into the oceans, creating enormous, ice-shelves, half a mile thick.

Life flourishes despite these severe conditions, and the terrain serves as a living laboratory for teams of scientists from around the world. The lichens growing on the rocks of the cold, dry valleys of Victoria Land, for example, give vital clues to life in severe conditions, such as those millions of years ago on Mars, and studies of the relationship between populations of natural predators and their prey in the seas surrounding the continent inform the management of fisheries worldwide.

Argentina, Australia, Chile, France, New Zealand, Norway, and the UK have all laid claim to parts of the Antarctic, although their sovereignty is not recognized by most other nations. The 1959 Antarctic Treaty promotes scientific co-operation and prohibits military activity, such as weapons testing, and waste dumping. Participating countries designate Specially Protected Areas (see map), to which access is restricted in order to leave important wildlife features undisturbed. Until the mid-1980s, scientific bases created high levels of local contamination (waste products, oil and rubbish). More recently, the scientific community has realized the importance of maintaining the Antarctic in its pristine state, as far as possible. Unfortunately, this has coincided with an increase in tourism in the area. More than 30,000 tourists visited the continent in 2007–08 – four times as many as 10 years previously. This increases the risk of the introduction of invasive species of plants and animals, and of a cruise ship grounding in the treacherous waters and causing an oil spill.

A further, and potentially even more damaging, threat to the region is posed by climate change. Temperatures on the Antarctic Peninsula have risen by around 2.5°C since the 1950s, leading in recent years to the dramatic disintegration of several ice shelves. The effect of this change on the flora and fauna is already being recorded, with the loss of sea-ice around the peninsula decimating the population of Adélie penguins, which use it to reach their feeding grounds. As their numbers dwindle, they are being replaced by gentoo penguins, which thrive in open waters. Warming ocean temperatures could also herald the arrival of a range of predatory species, such as king crabs, that could destroy the delicate ecosystem.

Krill – small crustaceans that provide food for fish, seabirds and mammals – feed on algae that form under sea-ice, and so are affected by its diminishing area. A more immediate threat, however, comes from commercial fishing, which began in the 1970s. Fears of over-exploitation led, in 1981, to the Convention on the Conservation of Antarctic Marine Living Resources, which regulates the annual krill catch but, despite this, krill stocks in 2003 were estimated at just one-fifth of their level in the 1970s, and the annual catch has continued to rise: from 109,000 tons in 2006–07 to over 684,000 tons the following season. It is feared that falling global fish-stocks and improved technology for catching krill could increase demand to several million tons a year, which would have serious implications for the survival of whales, seals, penguins and other birds.

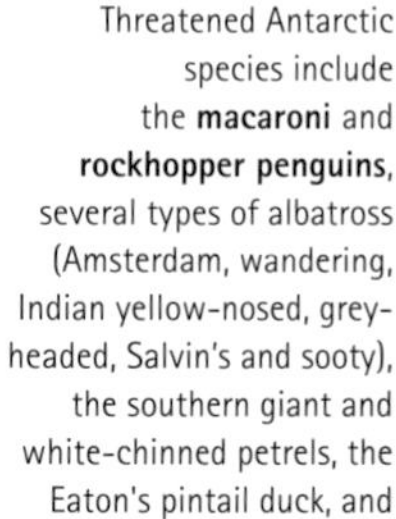

Threatened Antarctic species include the **macaroni** and **rockhopper penguins**, several types of albatross (Amsterdam, wandering, Indian yellow-nosed, grey-headed, Salvin's and sooty), the southern giant and white-chinned petrels, the Eaton's pintail duck, and the blue and fin whales.

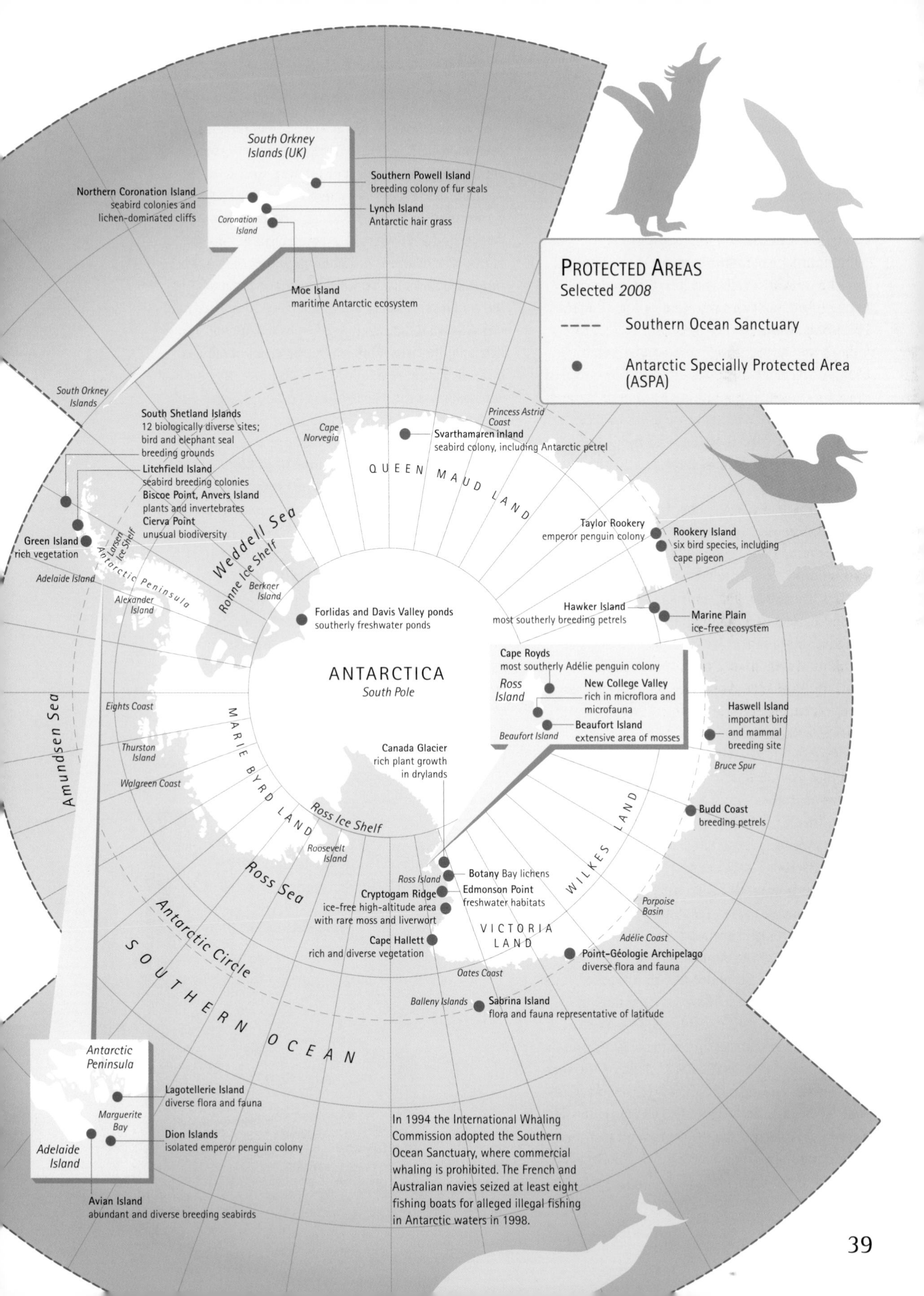

In 1994 the International Whaling Commission adopted the Southern Ocean Sanctuary, where commercial whaling is prohibited. The French and Australian navies seized at least eight fishing boats for alleged illegal fishing in Antarctic waters in 1998.

AUSTRALIA

At 3 million square miles (8 million sq km), Australia is the world's smallest continent. Yet its terrain is incredibly diverse, ranging from its central deserts to the rainforests of Queensland and Tasmania. The unique plant and animal life found in Australia is a reflection of its geographical isolation. Marsupials, or pouched mammals, such as the kangaroo or koala, have evolved into species as diverse as foetal mammals elsewhere in the world.

In an attempt to preserve Australia's rich biodiversity Unesco has designated a number of World Heritage Sites. Although some of these are recognized for their fossils, most have been awarded their special status because of the need to protect unique natural habitats.

Australia's flora and fauna are sensitive both to climate changes and to those brought about by humans. Some 34 species of animal are known to have become extinct in the past 30,000 years. The extinctions prior to the arrival of European settlers in the late 18th century were largely caused by an inadequate and fluctuating water supplies. More recent extinctions – of the Tasmanian tiger and several species of wallaby and bandicoot, for example – have been caused by humans.

Since their arrival in the late 18th century, European settlers have cleared and cultivated land in Australia, resulting in the loss of much of the country's woodland and rainforest. Tree felling has led to soil erosion, and to the water table rising, which increases the salinity of the soil and destroys any trees left standing. New trees have been introduced but this has not helped: some of these cause further damage. The native pastures of Queensland, for example, are threatened by the non-native prickly acacia.

Animal species that have been introduced have also wreaked havoc. The natural habitats of indigenous birds and marsupials have been disturbed by grazing cattle and sheep. Aggressive hunters such as the fox (introduced by European sportsmen as quarry) kill indigenous species, as well as the non-native rabbit. The cane toad, introduced as a pest-control measure, has itself turned into a pest.

Australia's rich biodiversity is, however, protected by a network of nearly 9,000 protected areas, covering 11 percent of the continent. The country also has a "necklace" of more than a dozen marine protected areas. UNESCO has designated a number of World Heritage Sites, many of which are intended to protect unique natural habitats. They include islands in the Indian Ocean and Tasman Sea for which Australia has responsibility. The most famous site is perhaps the Great Barrier Reef – 135,100 square miles (350,000 sq km) of coral. As one of Australia's major tourist attractions, government agencies and environmental groups have to work hard to protect it from the ravages inflicted each year by hundreds of thousands of visitors.

Under the 1999 Environment Protection and Biodiversity Conservation Act, a comprehensive national protection scheme for wildlife has been developed that includes rigorous assessment of building developments likely to affect the habitats of threatened or migratory species. More recently, in response to habitat changes brought about by global warming, the government expressed the intention to create a "climate corridor" of linked reserves and natural habitats, running the length of the continent's east coast, to enable species to move in response to changes in temperature and weather patterns. Time will tell whether these measures will reverse the decline of Australia's wildlife.

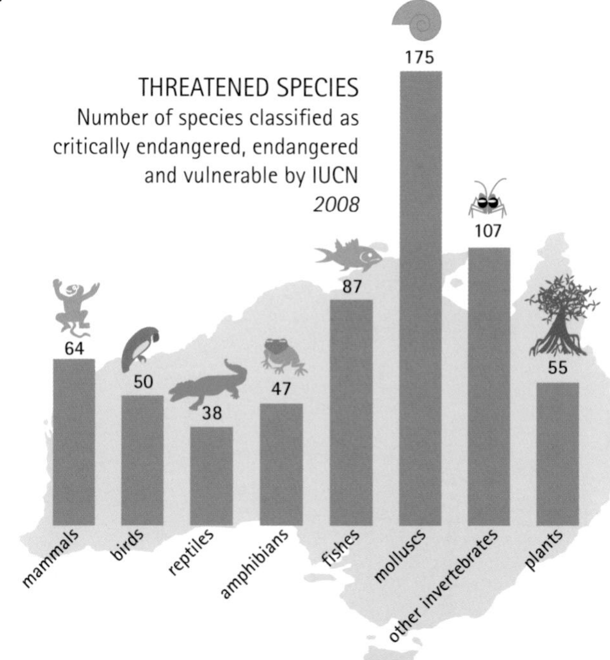

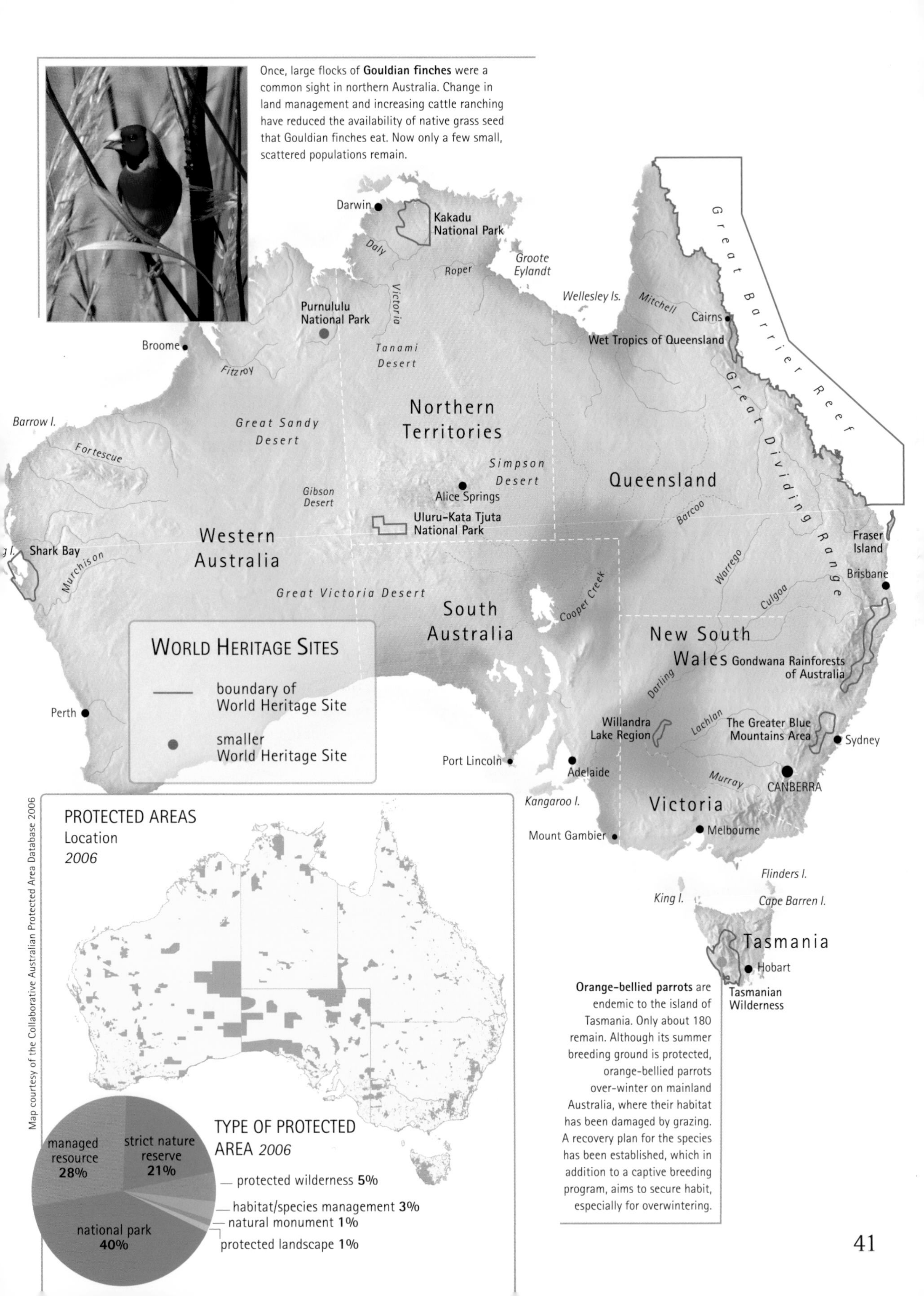

Once, large flocks of **Gouldian finches** were a common sight in northern Australia. Change in land management and increasing cattle ranching have reduced the availability of native grass seed that Gouldian finches eat. Now only a few small, scattered populations remain.

Orange-bellied parrots are endemic to the island of Tasmania. Only about 180 remain. Although its summer breeding ground is protected, orange-bellied parrots over-winter on mainland Australia, where their habitat has been damaged by grazing. A recovery plan for the species has been established, which in addition to a captive breeding program, aims to secure habit, especially for overwintering.

Central and South America

Central and South America is the most biologically diverse continent of the world. It includes temperate and tropical forests, high-altitude desert plateau, glacial peaks, freshwater and saltwater wetlands and coral reefs. While some areas remain relatively untouched by humans, others have been decimated by rapid urban growth, large-scale agriculture and mining.

The largest and best-known area under threat is the Amazon rainforest, home to several million species of plants, animals (including 3,000 species of fish) and micro-organisms, many still unrecorded by science. Many of these species are confined to small areas and are sensitive to environmental change so are at a high risk of becoming extinct with the destruction of their habitat.

The region's indigenous peoples historically lived by hunting, gathering and subsistence farming, mostly in small isolated tribes. Cattle ranching, mining and the timber industry were brought to the region by Europeans. Initially, settlement was restricted to the banks of the navigable rivers, still the principal means of transport, but roads are increasingly being built across the forest, some of them by the government, but most of them illegally, by developers.

The extent of the loss of rainforest habitat is difficult to assess in this huge, impenetrable area. The Brazilian National Institute of Space Research estimated deforestation between 2003 and 2004 to be 10,500 square miles (27,300 sq km). Although the annual rate declined in subsequent years, there were worrying signs, early in 2008, of an increase. In addition, mining activities and the effluent it produces, which can contain mercury, has severely polluted parts of the Amazon River.

Attempts are being made by governments, non-governmental organizations such as the WWF, Conservation International, university teams around the world, and wealthy individuals to protect parts of the remaining forest. One such scheme, the Amazon Region Protected Areas (ARPA) program, was set up in 2003 by WWF, the World Bank and the German Development Bank in collaboration with the Brazilian government, It aims to establish 109,000 square miles (283,000 sq km) of new protected areas and transform existing but neglected parks through a system of well-managed protected areas and sustainably managed reserves. At the same time as the Brazilian government is supporting such measures, however, it is pressing ahead with an infrastructure program that includes the development of roads through the rainforest, railways and dams, all of which will increase the rate at which vital habitat is lost.

The financial value of the rainforest is huge. Not only is it a source of potentially life-saving natural products that can be used in pharmaceuticals, but its vital function as a huge carbon store gives Brazil the opportunity to sell "carbon credits" to organizations unwilling to reduce their emissions of carbon dioxide. Its ecological value is immeasurable, and its loss could lead to a catastrophic increase in the rate of climate change.

Important ecological areas do not fit neatly within political boundaries, and conservation projects are increasingly seeking co-operation between countries. In Central America, for example, where there are more than 400 protected areas, efforts are being made to establish a "Mesoamerican biological corridor" that integrates the efforts of eight countries. And although other forest and mountain areas are now fairly well protected, it is recognized that more needs to be done to conserve the region's wetlands, such as the Pantanal, and its coastal and marine areas. Natural World Heritage Sites have been established throughout the continent, and include a wide range of habitats. International co-operation is not always so forthcoming. Attempts to protect the biologically rich mountainous Cordillera del Condor region in South America, for example, have been hampered by a border dispute between Ecuador and Peru

Protected Land

Percentage of total land area protected under IUCN categories I-V *2006*

- 24% or more
- 11% - 19%
- 1% - 9%
- less than 1%
- none or no data
- World Heritage Site
- World Heritage Site on danger list in *2007*

The Sea of Cortez World Heritage Site comprises 224 islands, islets and coastal areas, home to 695 plants, nearly 900 fish species, and nearly two-fifths of all species of marine mammals, including the **California sea lion**, and a third of all cetacean species.

Whale Sanctuary of El Vizcaíno
Sea of Cortez and Islands
MEXICO
BAHAMAS
TURKS & CAICOS
CAYMAN IS.
CUBA
Alejandro de Humboldt National Park
Sian Ka'an
Barrier-Reef Reserve System
DOMINICAN REP.
PUERTO RICO
Desembarco del Granma National Park
HAITI
VIRGIN IS. (UK)
Tikal National Park
BELIZE
Río Platano Biosphere Reserve
ST KITTS & NEVIS
GUADELOUPE
DOMINICA
GUATEMALA
HONDURAS
MARTINIQUE
ST LUCIA
Pitons
Morne Trois Pitons National Park
NICARAGUA
ST VINCENT & GRENADINES
BARBADOS
COSTA RICA
PANAMA
TRINIDAD & TOBAGO
Area De Conservación Guanacaste
Darien National Park
Coiba National Park
VENEZUELA
GUYANA
Los Katios National Park
FRENCH GUIANA
COCOS IS.
MALPELO IS.
COLOMBIA
SURINAME
Cocos Island National Park
Canaima National Park
Malpelo Fauna and Flora Sanctuary
Central Suriname Nature Reserve
Galapagos Islands
GALAPAGOS IS.
ECUADOR
Jaú National Park
Sangay National Park
PERU
Central Amazon Conservation Complex
Huascarán National Park
BRAZIL
Manu National Park
BOLIVIA
Discovery Coast Atlantic Forest Reserves
Noel Kempff Mercado National Park
Pantanal Conservation Area
Atlantic Forest Southeast Reserves
PARAGUAY
Iguaçu National Park
Iguazu National Park
CHILE
URUGUAY
ARGENTINA
Península Valdés
Los Glaciares
FALKLAND IS. (MALVINAS)

The Amazon rainforest is being lost at an unsustainable rate, chopped down mainly to create land for agriculture. A commitment by the Brazilian government in 2007 to complete the paving of the BR-163 north–south road through the forest will undoubtedly facilitate this process. In addition, a network of **unofficial roads**, created by logging and mining companies, is spreading throughout the forest.

The world's largest freshwater wetland, the **Pantanal** is shared by Bolivia, Brazil and Paraguay. It provides habitat for threatened species such as the giant river otter and the marsh deer, along with 120 other mammal species, 650 birds, 90 reptiles and 40 amphibians. Much of it is used by ranchers, and conservationists are working with them to ensure that the area is developed sustainably.

The Atlantic Forest has been reduced to 8% its original size, and is dangerously fragmented, putting at grave risk the survival of the **muriqui** and 20 other mammals found nowhere else in the world. Its rich diversity includes 2,200 species of birds and 20,000 species of plants, many of which would be lost if the forest were degraded further.

GALAPAGOS ISLANDS

The Galapagos Islands lie in the Pacific Ocean, about 625 miles (1,000 km) off the coast of South America. Because these volcanic islands have risen out of the seabed, evolution has taken place in isolation from the mainland, giving rise to species endemic to the islands. As Charles Darwin realized after his visit to the islands in 1835, species on neighbouring islands within the group have evolved unique characteristics in response to their environment. The result is a number of species and subspecies with small, localized populations, leaving them vulnerable to habitat destruction, disease, or climate change.

"Galapagos" means "tortoise" in Spanish, and the islands are home to giant tortoises, as well as about 300,000 marine iguanas. In the absence of indigenous predators – rice rats and two species of bat are the only native mammals – seabirds have bred freely, and large colonies of boobies, frigate birds, and the rare lava gull have flourished. In 1959, Ecuador declared the islands a national park, and in 1978 they became the first United Nations World Heritage Site. In 2007, however, they were placed on the World Heritage in Danger list because of the threat to their unique biodiversity from invasive species, tourism and immigration.

Only some of the islands have been settled, but the total population has grown from 2,000 in 1970 to 18,000 in 2008. Settlers have so far brought about 500 species of plants to the Galapagos, which are displacing the native flora. Introduced animal species include rats, cats, and goats (successfully eradicated from some islands). Tourism is the mainstay of the economy, with over 108,000 visitors in 2004, but fishing is also important. At least 800 fishermen are based on the islands, and larger boats fish off-shore, creating a hazard to marine mammals and birds.

In April 1997, the President of Ecuador issued an emergency decree, restricting the introduction of alien species, and promoting conservation. In 1998, the no-fishing zone around the islands was extended from 15 to 40 nautical miles, creating a marine reserve of over 50,000 square miles (130,000 sq km). However, attempts to limit fishing have been met with resistance from local people. The fate of the wildlife on the islands hangs on whether both economic and conservation needs can be met.

There are only around 1,300 **Galapagos penguins**, and their population appears to fluctuate according to the availability of food. Their fish diet relies on cold, nutrient-rich waters, but the El Niño phenomenon causes a flow of warmer waters, which disrupts their supply. The more frequent El Niño episodes, predicted as a result of climate change, could place them under greater threat.

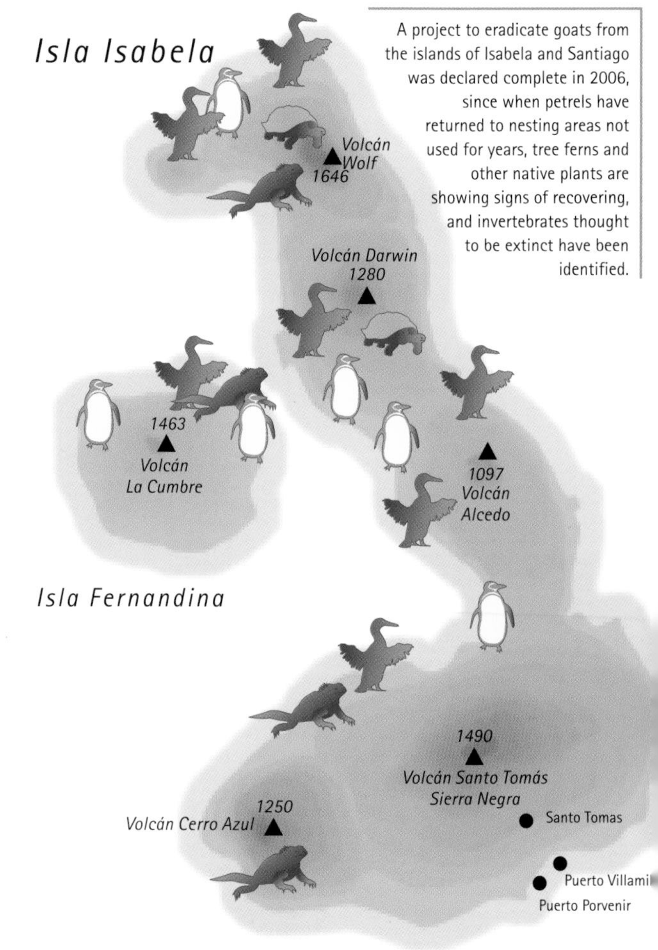

A project to eradicate goats from the islands of Isabela and Santiago was declared complete in 2006, since when petrels have returned to nesting areas not used for years, tree ferns and other native plants are showing signs of recovering, and invertebrates thought to be extinct have been identified.

Found only on the Galapagos, the **lava gull**, with 300 to 400 breeding pairs, is considered the rarest gull in the world.

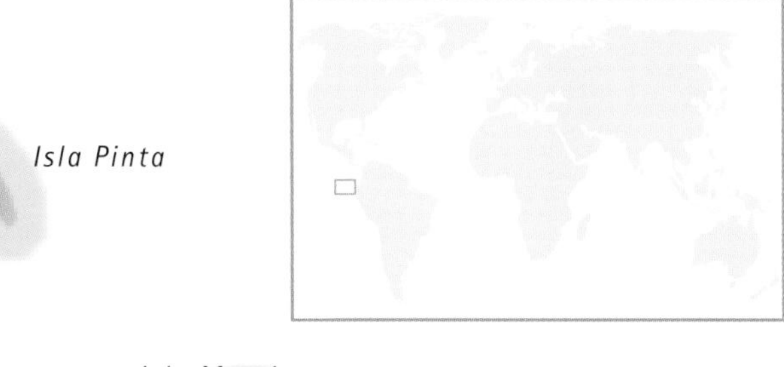

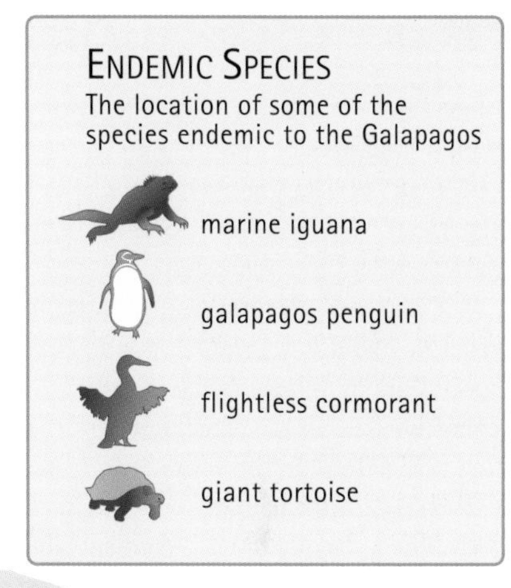

Isla Pinta

Isla Marchena

Isla Genovesa

The **giant tortoise**, once numerous on the islands, presented an easy meal to passing sailors and three of the 14 known subspecies became extinct. In recent times, the main threat to the tortoise was the domestic goat, which competes with it for grazing, but this has now been eradicated from Pinta. This has come too late for "Lonesome George", however, who is the sole survivor. There are plans to introduce captive-bred Española tortoises to the island.

GIANT TORTOISE
Estimated number of individuals in *1535* and *2008*

250,000 — *1535*

15,000 — *2008*

Isla Santiago

Isla Rábida

Isla Santa Cruz

Isla Pinzón

Since 1970 more than 500 young tortoises, bred in captivity, have been released on the island of Pinzon.

Puerto Ayora

Isla Santa Fé

Between the mid-1960s and 1974, 14 giant tortoises, whose grazing had been decimated by goats, were taken from Española to breeding pens on Santa Cruz island. The goats were eventually removed from the island and by March 2000 a total of 1,000 tortoises had been reintroduced to the island.

On 16 January 2001, the oil tanker *Jessica*, carrying 160,000 gallons of diesel and 80,000 gallons of bunker fuel, ran aground off San Cristóbal Island. Her tanks ruptured. Fortunately, winds and currents took the oil away from the island, preventing a major ecological disaster.

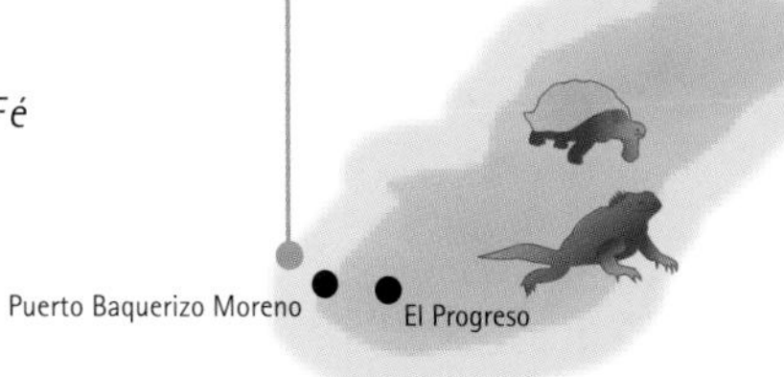

Puerto Baquerizo Moreno

El Progreso

Isla San Cristóbal

Puerto Velasco Ibarra

Isla Santa María

Isla Española

MADAGASCAR

Madagascar is the world's fourth largest island, with a land area of approximately 226,000 square miles (585,000 sq km). It is thought to have separated from Gondwanaland (present-day Africa) 60 million years ago.

Madagascar's geographical isolation has led to separate evolution, producing many endemic species and a diverse range of habitat: tropical conditions along the coast, temperate inland areas and arid deserts in the south.

Several of Madagascar's endemic species are threatened, despite the government's endorsement of international agreements on biodiversity, desertification, endangered species, and marine life conservation. With a human population exceeding 20 million and growing at an annual rate of nearly 3 percent, the pressure on the country's natural habitats is likely to continue.

Although the great elephantbird and Delalande's coua, or snail-eating coua, are known to have become extinct, new species are being discovered all the time. Lemurs are primates that were displaced by monkeys elsewhere in the world and are now found only on Madagascar. A new species of woolly lemur, five new dwarf lemurs, and eight new species of mouse lemur have been announced since the beginning of the century.

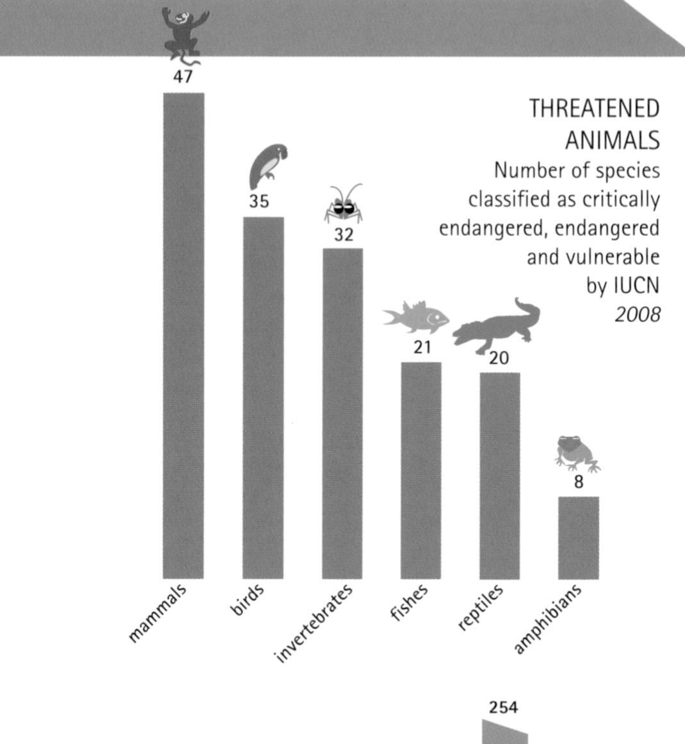

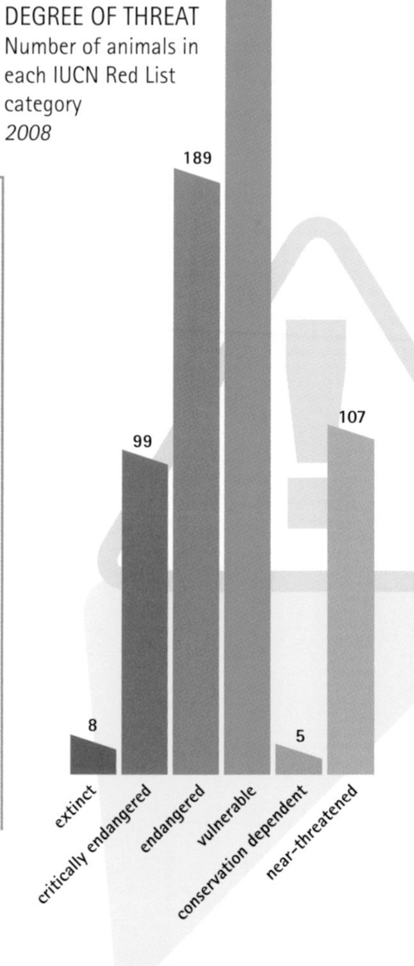

The **Madagascar rosy periwinkle** is the source of two alkaloids – vincristine and vinblastine – used to treat (and usually cure) childhood leukaemia. The plants from which the alkaloids are extracted are now grown elsewhere. The drug industry has not returned any benefit to the people or to conservation projects in Madagascar.

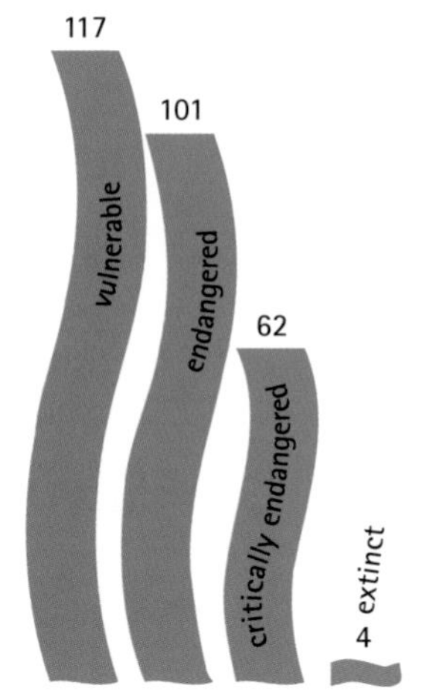

The **fish eagle** was common in north-west Madagascar until the 1930s. It is now very rare, as a result of persecution, hunting, and the loss of nesting and feeding habitat.

Tsingy de Bemaraha Strict Nature Reserve is a World Heritage Site. Endangered lemurs and birds thrive in the undisturbed forests, lakes and mangrove swamps.

In 2007, the **Rainforests of the Atsinanana**, comprising a string of national parks down the eastern side of the island, was designated a World Heritage Site.

The **golden lemur** is found only in the southeast of the island. About 1,000 inhabit the Ranomafana National Park. They eat the base of bamboo, and the new shoots, which are lethally toxic to most other mammals.

Endangered Animals and Plants

"The time is past when large, rare creatures can recover their numbers without man's strenuous intervention."

– Peter Matthiessen,
author and naturalist

PRIMATES

Apes, humans, monkeys, and prosimians such as lorises, bush babies, and lemurs make up the 234 species of primates, a mammalian order with distinctive common features indicating descent from a single ancestor: a small, tree-dwelling mammal that subsisted on fruit and insects.

Primates have prehensile hands with opposable thumbs, large brains (especially the cerebral cortex), and usually bear single offspring. Most are well adapted to life in trees, especially tropical forests, although some species have become terrestrial. Primates tend to live in complex social groups, and as infants are dependent on their mother, both for sustanance and for lessons in the practical and social skills. With a few exceptions, such as the Japanese macaque that lives on the Shimokita Peninsula at a latitude of 41°N, very few primates have colonized temperate areas because they need a supply of food during the winter months as well as long daylight hours for foraging.

All primates share behavioral and anatomical characteristics, but humans and chimpanzees are particularly close. Chimps share almost 99 percent of our DNA. They use tools, laugh when tickled, and – when allowed – can live for 60 years. Like many other species of primate, chimps are endangered by deforestation and, especially, its consequences: as logging companies open up the forests, hunters move in. Since the mid-1980s hunting for "bushmeat" has become a lucrative industry. In 2001, over a million tonnes was taken from the Congo basin alone. Living ever-closer to humans, primates also suffer from diseases such as the Ebola virus.

A hundred years ago there were some two million chimps living in the Central African rainforest, stretching from Sierra Leone to Tanzania. By 2001, only 200,000 remained and their numbers continue to dwindle.

About 90 percent of **primates** live in tropical forests. They play an integral role in the ecology of their habitat, helping to pollinate plants and disperse seeds.

Our closest relatives are the four great apes: **gorilla**, chimpanzee, bonobo (pygmy chimpanzee), and orang-utan.

Only 5% of Cameroon's original forest remains. It is inhabited by the critically endangered lowland gorilla. The Cameroon government uses armed patrols to protect gorilla colonies from poachers. In 2007, they also negotiated the return of four gorillas who had been captured five years previously and held in zoos in Malaysia and South Africa.

In the late 1990s, over 150 rare **eastern lowland gorillas** were reported to have been casualties of the civil unrest in Rwanda.

The **bonobo**, or pygmy chimp, has been a victim of the war in the Democratic Republic of Congo. Fewer than 50,000 remain.

Lemurs are found only on Madagascar, where they have evolved in isolation for 100 million years. They range from the size of a mouse, to the size of a panda. A third of lemur species are thought to have become extinct since 1500 and, of the remainder, four-fifths are under threat from habitat loss and human hunting.

Many endangered primates live in countries least able to sustain them, and suffer the consequences of poverty, ecological degradation, and conflict. The demand for palm oil and biodiesel drives the conversion of tropical forest to plantations in Indonesia, a dire threat to the remaining **orang-utan**.

CATS

Cats belong to the Felidae family – a group of carnivorous mammals that includes the lion, tiger, jaguar, leopard, puma, lynx and the domestic cat.

Cats, whether large or small, are built for performance and are more specialized in this respect than any other flesh-eating mammal. They are powerfully built animals, with large and highly developed brains, making them more intelligent as well as stronger than their prey. They are also so well-coordinated that they almost always land on their feet when they fall.

Although the lion, tiger, and cheetah are agile climbers, they are mainly terrestrial in habit. The leopard, jaguar, and ocelot, in contrast, are very much at home in trees, where they sometimes sleep. The larger cats range over wide areas, often up to 50 square miles (129 sq km). Big cats usually rove alone or with a companion. African lions do form prides but these loosely bound groups of females, cubs and one or a few adult males lack the rigid hierarchy found in dog and wolf packs.

Humans are the main threat to big cats and hunting has long been a popular pastime. It remains a lucrative sport: in Botswana, for example, recreational hunters may pay up to US $30,000 to shoot a lion. The skin, teeth and claws of big cats provide rich pickings for traders. Those not threatened by hunters have had their extensive territories steadily eroded by agriculture and human settlement.

Many of Asia's big cats are under threat, including the snow leopard (reduced to around 6,000 in the wild), the Siberian tiger (reduced to 400 in the wild) and the Amur leopard (see page 95). As wild populations of big cats decline towards extinction, zoos and other sanctuaries become even more vital in ensuring their survival.

The highly endangered **Florida panther** is making its last stand in the higher elevations in the swamps of the Everglades. The few remaining animals became highly inbred, causing such genetic flaws as heart defects and sterility. In the 1990s, closely related panthers from Texas were released in Florida and are successfully breeding with the Florida panthers. Increased genetic variation and protection of habitat may yet save this subspecies.

The **cheetah** is generally considered the speediest of animals, capable of 60-70 mph (100-110 kmph), with rare reports of even higher speeds. A cheetah may also pursue its prey for a considerable distance, as much as 3.5 miles (5.5 km). But even its speed may not save it from hunters and the threat of extinction.

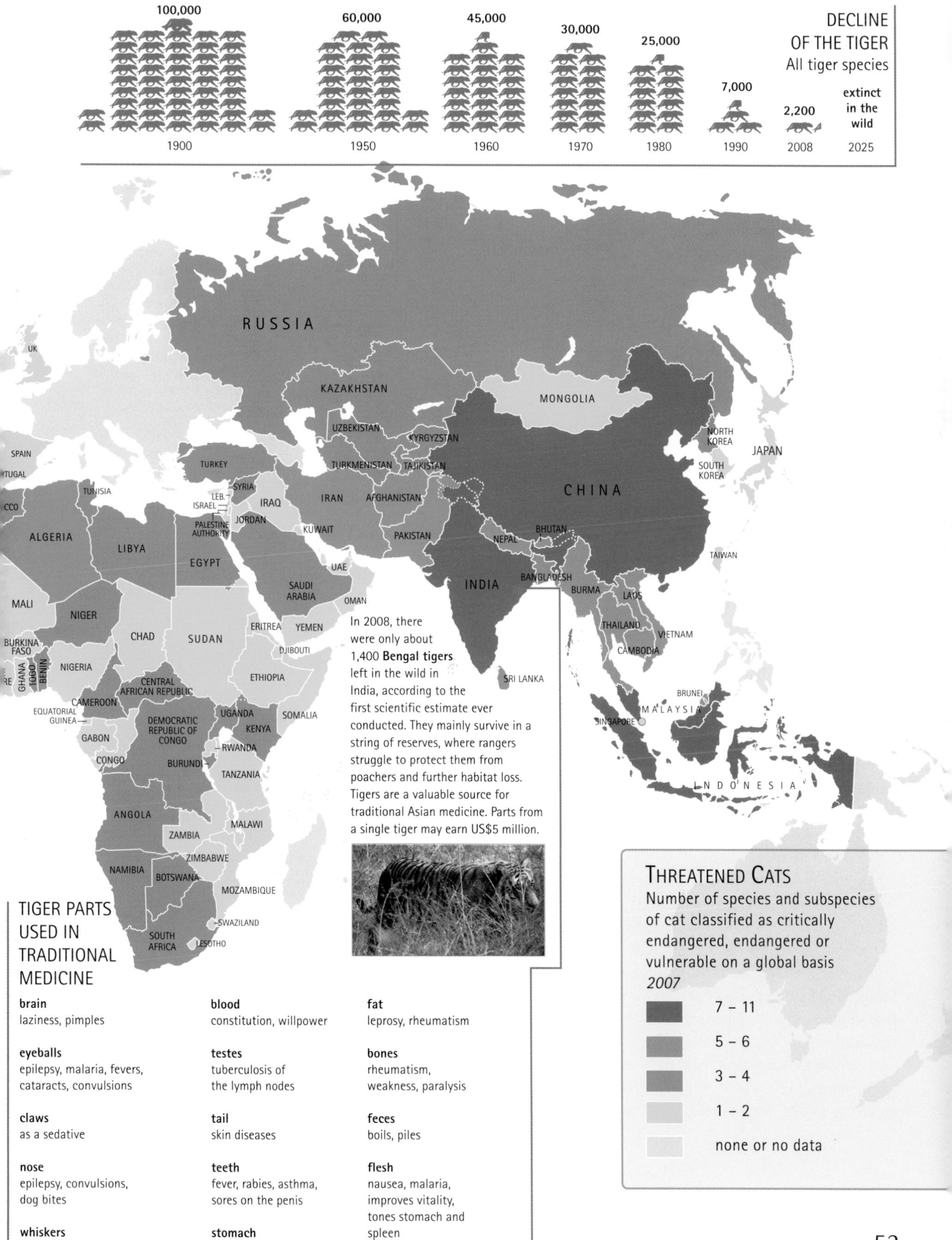

DECLINE OF THE TIGER
All tiger species
100,000
1900
60,000
1950
45,000
1960
30,000
1970
25,000
1980
7,000
1990
2,200
2008
extinct in the wild
2025
RUSSIA
UK
KAZAKHSTAN
MONGOLIA
UZBEKISTAN
KYRGYZSTAN
NORTH KOREA
SOUTH KOREA
JAPAN
SPAIN
TUGAL
TURKEY
TURKMENISTAN
TAJIKISTAN
CHINA
CCO
TUNISIA
SYRIA
LEB.
ISRAEL
IRAQ
IRAN
AFGHANISTAN
PALESTINE AUTHORITY
JORDAN
KUWAIT
ALGERIA
LIBYA
EGYPT
PAKISTAN
NEPAL
BHUTAN
TAIWAN
UAE
SAUDI ARABIA
INDIA
BANGLADESH
BURMA
LAOS
MALI
NIGER
OMAN
CHAD
SUDAN
ERITREA
YEMEN
THAILAND
VIETNAM
BURKINA FASO
DJIBOUTI
CAMBODIA
GHANA
TOGO
BENIN
NIGERIA
ETHIOPIA
SRI LANKA
CENTRAL AFRICAN REPUBLIC
CAMEROON
BRUNEI
EQUATORIAL GUINEA
UGANDA
SOMALIA
MALAYSIA
DEMOCRATIC REPUBLIC OF CONGO
KENYA
SINGAPORE
GABON
RWANDA
CONGO
BURUNDI
TANZANIA
INDONESIA
In 2008, there were only about 1,400 **Bengal tigers** left in the wild in India, according to the first scientific estimate ever conducted. They mainly survive in a string of reserves, where rangers struggle to protect them from poachers and further habitat loss. Tigers are a valuable source for traditional Asian medicine. Parts from a single tiger may earn US$5 million.
ANGOLA
MALAWI
ZAMBIA
ZIMBABWE
NAMIBIA
BOTSWANA
MOZAMBIQUE
SWAZILAND
SOUTH AFRICA
LESOTHO
THREATENED CATS
Number of species and subspecies of cat classified as critically endangered, endangered or vulnerable on a global basis
2007
7 – 11
5 – 6
3 – 4
1 – 2
none or no data
TIGER PARTS USED IN TRADITIONAL MEDICINE
brain
laziness, pimples
blood
constitution, willpower
fat
leprosy, rheumatism
eyeballs
epilepsy, malaria, fevers, cataracts, convulsions
testes
tuberculosis of the lymph nodes
bones
rheumatism, weakness, paralysis
claws
as a sedative
tail
skin diseases
feces
boils, piles
nose
epilepsy, convulsions, dog bites
teeth
fever, rabies, asthma, sores on the penis
flesh
nausea, malaria, improves vitality, tones stomach and spleen
whiskers
toothache
stomach
stomach upsets

UNGULATES

Horses, deer, cattle, pigs, sheep and goats are all ungulates – mammals with hooves of hard skin that allow the animals to run. Elephants and rhinoceroses (see pages 56-57) are also ungulates. Their teeth are adapted for their plant diet, with strong molars for grinding. Preyed upon by big cats, wolves and other carnivores, ungulates rely on various methods of defence. Some are large, many are swift, and others grow horns or antlers. They often mass in large herds to minimize the risk to an individual.

Domesticated ungulates accompanied earlier European explorers to Australia and to isolated islands all over the world, but have often threatened indigenous herbivores by over-grazing native plants. Non-native ungulates have interbred with indigenous species, and have also transmitted diseases such as rinderpest, which spreads from cattle to wild buffalo in Southeast Asia.

Ungulates provide an important source of meat for humans, and their hides are used for clothing and shelter. As vehicles and guns improved during the 20th century, ungulate populations began to suffer from the over-hunting that remains the most serious threat to their survival.

Areas designated as reserves are often vital for ungulate conservation. The tourists flocking to witness herds of ungulates roaming over African plains, provide a potential source of income for conservation in reserves. Kruger National Park in South Africa is Africa's oldest reserve, founded in 1898, while Tanzania boasts the Serengeti and Ngorongoro parks.

To secure local support for conservation inside and outside reserves local people must enjoy the profits from hunting and tourism and participate in the management of their wildlife. The Communal Area Management Programme for Indigenous Resources (CAMPFIRE) in Zimbabwe is a successful example, demonstrating that the managed re-introduction of ungulates to areas where they had been hunted to local extinction can be successful with proper enforcement against poaching.

There may have been 60 million **bison** in North America before the arrival of Europeans. Hunting drove the bison to the point of extinction as settlers swept west across the continent. About 1,500 survived in areas such as Yellowstone National Park, established in 1872. The number of bison has since risen to over 400,000, most reared commercially for meat, hide and other products.

CHANGING FORTUNES OF THE NORTH AMERICAN BISON

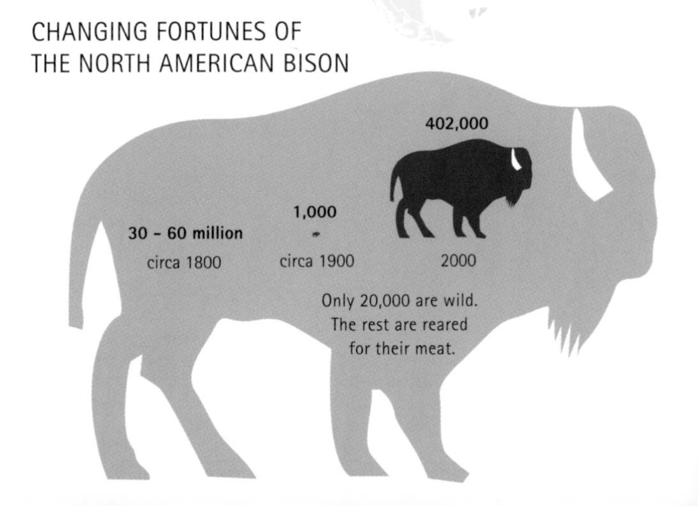

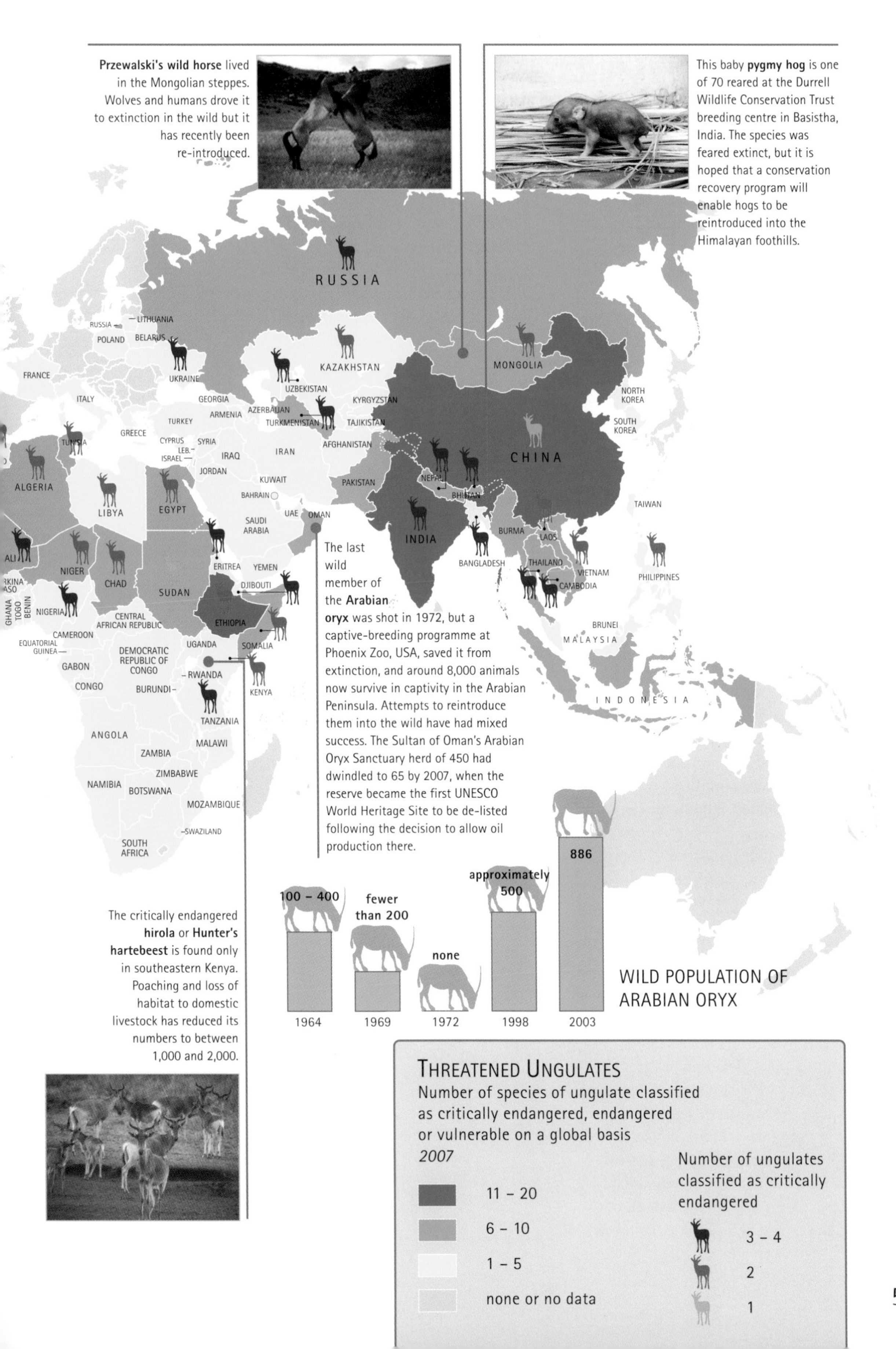

Przewalski's wild horse lived in the Mongolian steppes. Wolves and humans drove it to extinction in the wild but it has recently been re-introduced.

This baby **pygmy hog** is one of 70 reared at the Durrell Wildlife Conservation Trust breeding centre in Basistha, India. The species was feared extinct, but it is hoped that a conservation recovery program will enable hogs to be reintroduced into the Himalayan foothills.

The last wild member of the **Arabian oryx** was shot in 1972, but a captive-breeding programme at Phoenix Zoo, USA, saved it from extinction, and around 8,000 animals now survive in captivity in the Arabian Peninsula. Attempts to reintroduce them into the wild have had mixed success. The Sultan of Oman's Arabian Oryx Sanctuary herd of 450 had dwindled to 65 by 2007, when the reserve became the first UNESCO World Heritage Site to be de-listed following the decision to allow oil production there.

The critically endangered **hirola** or **Hunter's hartebeest** is found only in southeastern Kenya. Poaching and loss of habitat to domestic livestock has reduced its numbers to between 1,000 and 2,000.

Elephants and Rhinos

The elephant family comprises two species: the Indian elephant (*Elephas maximus*), weighing about 5.5 tonnes, and the African elephant (*Loxodonta africana*) weighing up to 7.5 tons and standing between 10 and 12 feet (3–4 m) at the shoulder. Both species live in habitats ranging from thick jungle to savanna. They form small family groups led by the eldest females, and where food is plentiful groups join to create larger herds. Most bulls live in bachelor herds apart from the cows. Elephants migrate seasonally, according to the availability of food and water. They spend many hours eating and may consume 500 lbs (225 kg) of grasses and other vegetation in a day.

For many centuries the Indian elephant has been used as both a ceremonial and draft animal. Elephants have been crucial to Southeast Asian logging operations, for example. African elephants are also used as working animals, but not as extensively.

Elephants are in great danger from habitat destruction and human exploitation. Both Indian and African elephants are classified as endangered; the African elephant suffers in particular from poaching for the ivory trade. Although conservation measures have been taken, including patrols to protect against poachers, and the creation of large reserves, these can lead to further problems. In good conditions, elephant populations can increase at a rate of 5 percent a year which leads to overpopulation and may make culling necessary. Corridor areas are vital to protect major migratory routes and to prevent herds from becoming isolated.

Rhinoceroses have one or two horns on the upper surface of the snout, composed of keratin, a fibrous protein found in hair. The Indian rhinoceros (*R. unicornis*) is the largest of the family, at about 14 feet (4.3 m) long and weighing up to 5 tons. Most rhinoceroses are solitary inhabitants of savanna, scrub forest, or marsh, although the Sumatran rhino is now found only in deep forest. Rhinoceroses have poor eyesight but acute hearing and sense of smell. Despite their bulk, they are remarkably agile; the black rhino may attain a speed of about 30 mph (45 kmph), even in thick brush.

All but the white rhinoceros are listed as endangered in the Red Data Book. Despite protective laws, poaching continues to supply a thriving market, with rhinoceros horn and blood among the most highly-prized ingredients in traditional medicines.

Captive-breeding programs offer the only hope for maintaining some species until adequate protection can be provided in the wild.

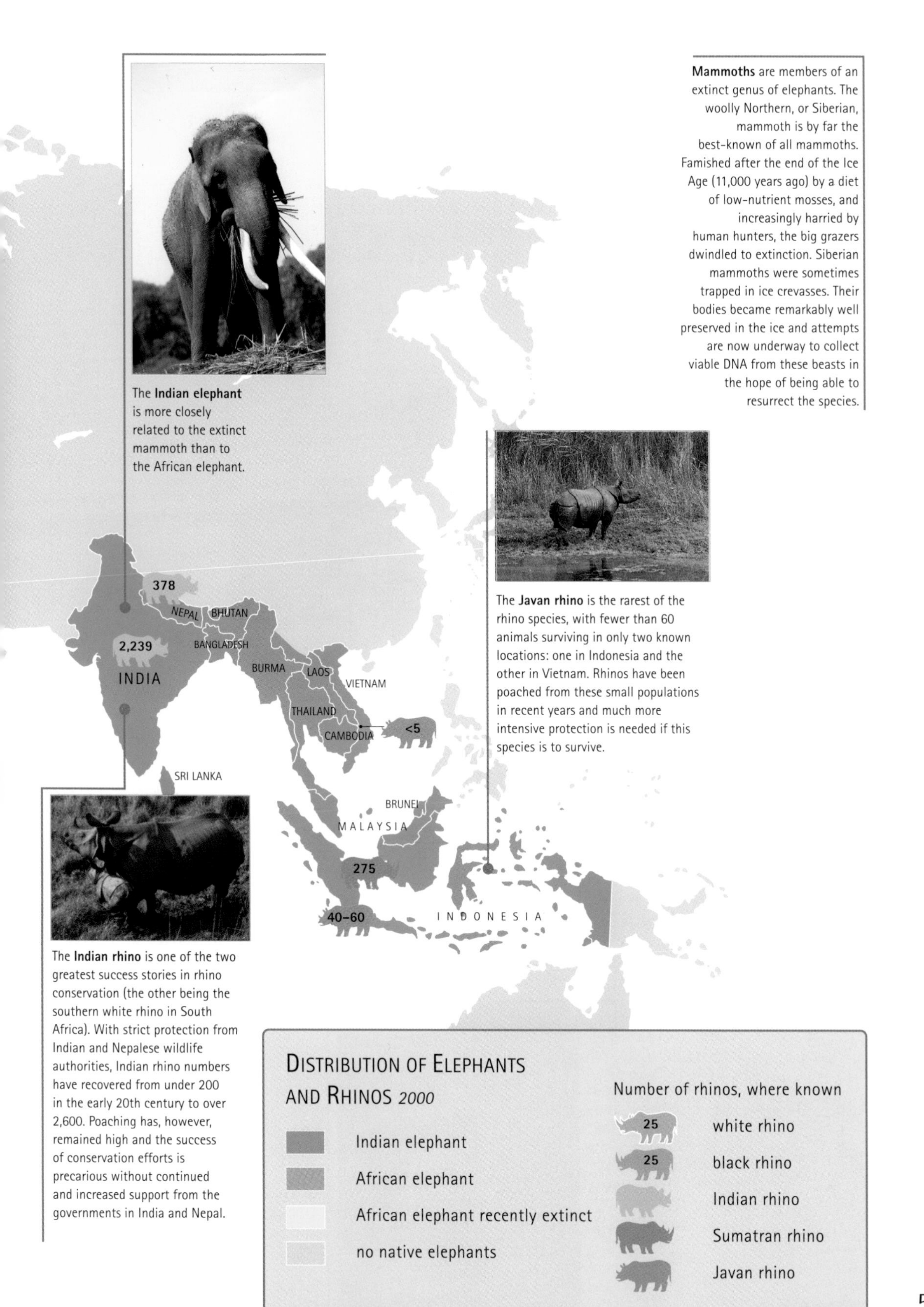

The **Indian elephant** is more closely related to the extinct mammoth than to the African elephant.

Mammoths are members of an extinct genus of elephants. The woolly Northern, or Siberian, mammoth is by far the best-known of all mammoths. Famished after the end of the Ice Age (11,000 years ago) by a diet of low-nutrient mosses, and increasingly harried by human hunters, the big grazers dwindled to extinction. Siberian mammoths were sometimes trapped in ice crevasses. Their bodies became remarkably well preserved in the ice and attempts are now underway to collect viable DNA from these beasts in the hope of being able to resurrect the species.

The **Javan rhino** is the rarest of the rhino species, with fewer than 60 animals surviving in only two known locations: one in Indonesia and the other in Vietnam. Rhinos have been poached from these small populations in recent years and much more intensive protection is needed if this species is to survive.

The **Indian rhino** is one of the two greatest success stories in rhino conservation (the other being the southern white rhino in South Africa). With strict protection from Indian and Nepalese wildlife authorities, Indian rhino numbers have recovered from under 200 in the early 20th century to over 2,600. Poaching has, however, remained high and the success of conservation efforts is precarious without continued and increased support from the governments in India and Nepal.

Bears

Bears are among the largest of the carnivores and belong to the family Ursidae. Eight species are recognized, and there are numerous subspecies. Six species are threatened, as is the lesser panda, which belongs to the family Ailuridae. Insufficient data exist about the sun or honey bear but this species is also feared to be threatened.

Bears vary in size from the sun bear, weighing only 66 lbs (30 kg), to the Kodiak brown bear, which can weigh 1,540 lbs (700 kg). Their diet varies as well. Polar bears are particularly fond of seals, panda bears eat mainly bamboo, and sloth bears prefer insects. Most bears, of course, eat honey.

Bears tend to live solitary lives, pairing only to mate. They are strong swimmers and smaller species are agile climbers.

Bears are hunted as trophies, for their hides, for meat and also from fear. When their territory overlaps with cultivated land, they are perceived as pests. Poachers kill bears for body parts and trap them for the pet trade or for use as performing animals. Bear gallbladders are used in traditional Chinese medicine to cure liver disease, cancer and other ailments.

In many parts of the world bears' habitats have been eroded and their populations fragmented, leading to the threat of local extinction. Small, insular groups of bears can also become weakened by genetic depletion.

Sophisticated management of hunting and habitat protection has sustained bear species living in North America and the Arctic, although pollution and global warming are serious problems (see pages 36-37). Resources for conservation in Asia and South America are not as plentiful, however, and a growing human population demands more land, which inevitably threaten bears. The establishment of wildlife "corridors" between reserves allows bears from different populations to mate and mix genes. Captive-breeding programs also offer hope for endangered bears. Over 100 giant pandas are kept in the world's zoos, but captivity impedes their breeding success rate as well as their life span.

The **brown bear**, called "grizzly" in North America, once ranged over all the northern continents. It has been hunted near to extinction in Europe, where only isolated populations totalling 13,000 remain in mountain ranges. Attempts to reintroduce bears into the Pyrénées seem doomed.

VENEZUELA
COLOMBIA
ECUADOR
PERU
BOLIVIA
ARGENTINA

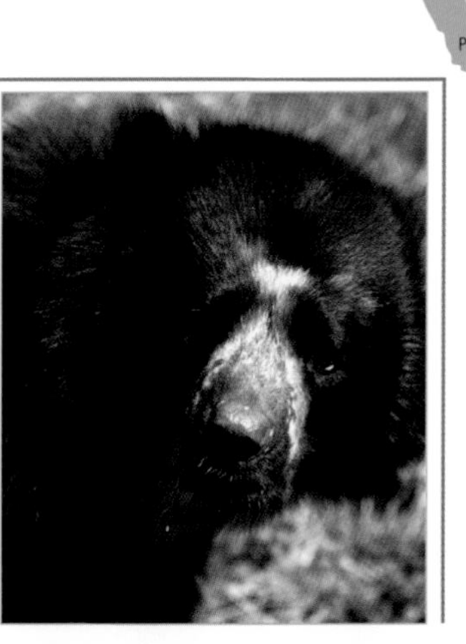

The **spectacled bear** is found in South America, living in terrains ranging from desert to rain forest, but it thrives best in cloud forest at around 2,000 metres. Logging and mining activites, and the expansion of agricultural land, is threatening the bear's survival. It is also hunted for its meat and skin. Attempts at conservation are hampered by unstable political environments and by drug trafficking.

In 2004 there were nearly 1,600 **giant pandas** left in the wild in the mountain bamboo forests in central China. The 50 state-run nature reserves shelter two-thirds of them. The rest face the threat of deforestation and poaching. The focus of conservation work is habitat protection, but there is also a captive-breeding programme, which produced 34 cubs in 2006. However, none has yet been introduced to the wild.

Threatened Pandas

Species of panda considered under threat *2000*

- lesser panda
- giant panda

The Baluchistan bear, a subspecies of the **Asiatic black bear**, is critically endangered.

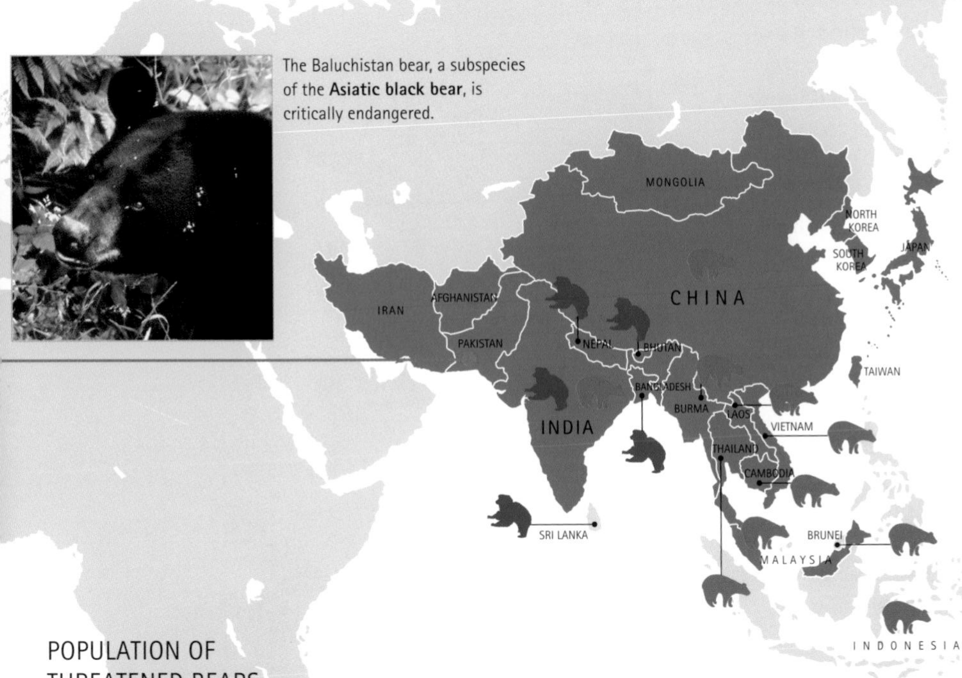

Population of Threatened Bears

2007 or latest available data

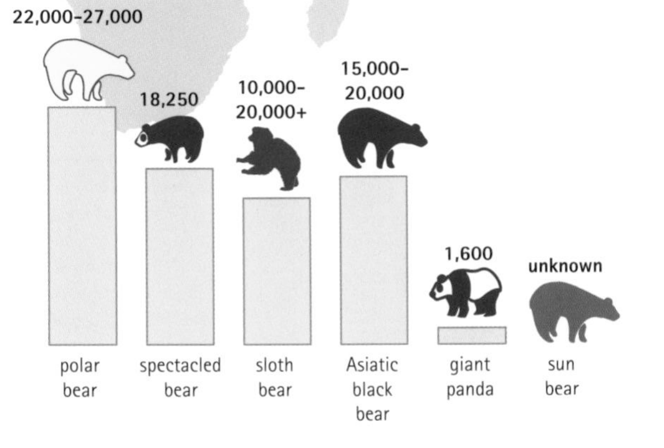

Threatened Bears

Species of bear considered under threat *2007*

- spectacled (Andean) bear
- Asiatic black bear
- sun (honey) bear
- sloth bear

Rodents

There are over 2,000 species of rodent, making them the most diverse order of mammals. They are found above and below ground and in all parts of the world except for New Zealand and the Antarctic. Most rodents are small – the dormouse weighs less than an ounce (20 grams) – although there are exceptions, such as the capybara of South America, which grows to 110 lbs (50 kg).

Rodents are commonly seen as agricultural pests, carriers of disease such as bubonic plague and as a threat to biodiversity. The black rat and Norway rat are cited as key villains, travelling all over the world via ships, wrecking food stores and displacing native rodent species. In fact, rodents are ecologically valuable. Some, such as squirrels, help plants to reproduce by burying seeds. Burrowing rodents mix, fertilize and aerate soil. As prey, rodents provide essential food for predators.

Rodent populations tend to be larger than those of other groups of mammals and their breeding rates are very high, making them well-equipped as a species to recover from adverse weather and epidemics. Despite this advantage, sustained pressure on rodent populations can cause local extermination; for species with only a limited range this can result in extinction.

The greatest threat to rodents is the progressive loss of their natural habitat from urbanization, cultivation of crops, and grazing of domestic livestock. Throughout the 20th century the US government sponsored a program of poisoning and plowing to eliminate prairie dogs, which degrade cattle-grazing land. As a result, the black-footed ferret, which preyed on the prairie dogs, disappeared. Although still classified as extinct in the wild, black-footed ferrets have been reintroduced with some success from a captive-bred population.

Fur trapping is illegal in many places, but continues to threaten some rodent species. Rarity and value tend to go hand in hand, thus encouraging trapping of the few remaining animals.

North American beavers dam rivers to create wetlands, providing habitat for a wide range of other animals. They were almost driven to extinction by fur trappers and land drainage in the early 20th century.

The population of the **short-tailed chinchilla** collapsed by 80% in the last decade of the 20th century due to depletion of its habitat in the mountains of South America, and illegal trapping for its valuable fur. It is now critically endangered.

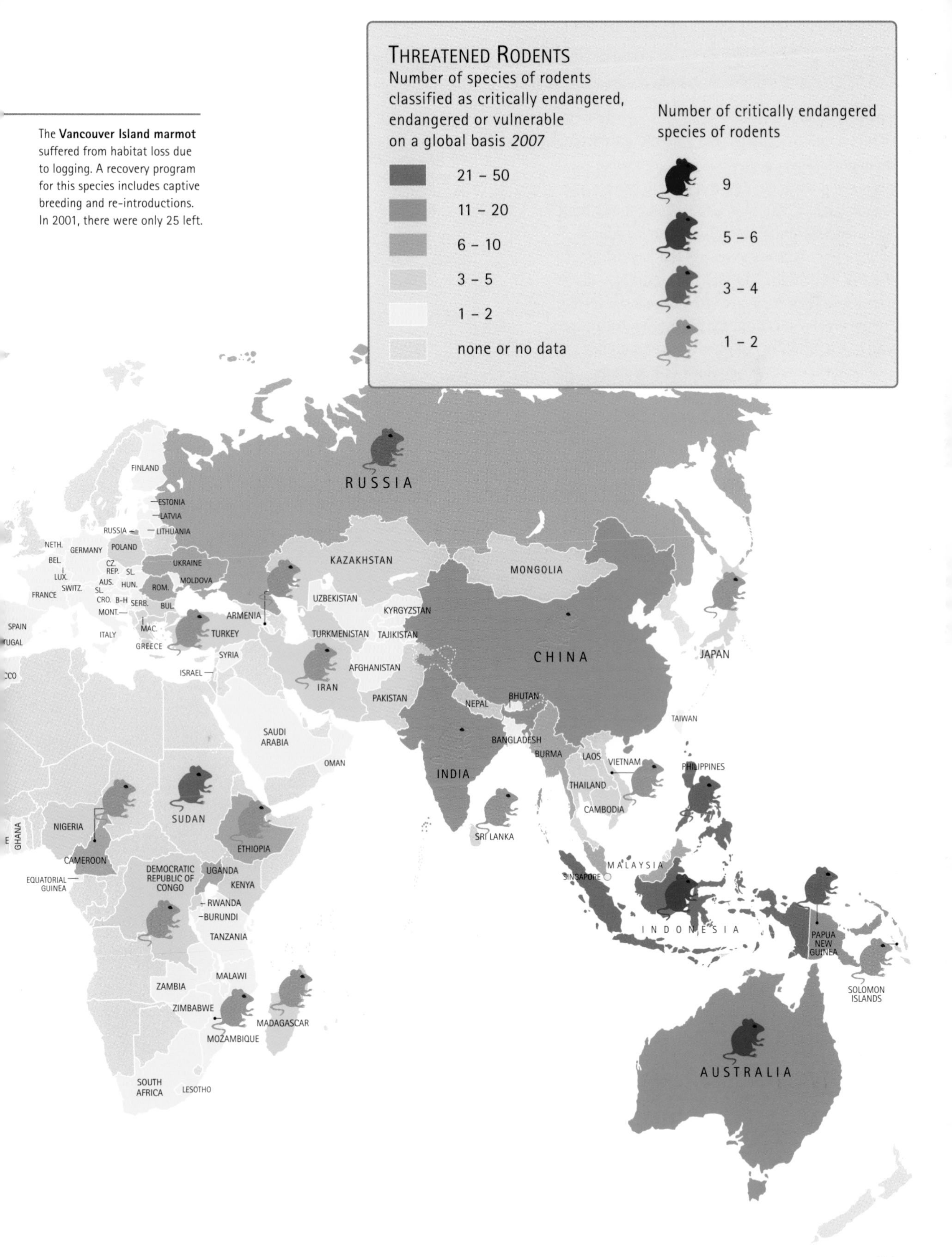

The **Vancouver Island marmot** suffered from habitat loss due to logging. A recovery program for this species includes captive breeding and re-introductions. In 2001, there were only 25 left.

Bats

There are about 900 species of bats, nearly a third of which are threatened. Bats are the only flying mammals. Most eat insects, although some eat nectar and fruit, helping to pollinate plants and disperse seeds. Vampire bats in Central and South America feed on the blood of large mammals.

Bats are nocturnal, avoiding the heat of the day, the risk of dehydration and the attentions of predators. They navigate at night using a form of sonar or echo-location. By day they roost in trees and caves, often in colonies of over a million bats. In urban areas they roost in buildings.

Bats are distributed over temperate and tropical regions but are most abundant near the equator. Those that spend the summer in temperate regions either hibernate or migrate towards the equator in winter. Most produce only one offspring each year, so bat populations tend to recover slowly from catastrophic events.

Bats bring several benefits to humans. Their droppings contain nitrogen and phosphorus and are collected for use as fertilizer. They also help to control populations of insects that are pests of crops and carry diseases. A single bat can kill 20,000 insects in one night. In contrast, they have only a few drawbacks for humans. Some commercial fruit farmers have to use nets to protect their crops from bats, and vampire bats can transmit diseases including rabies to their animal hosts.

Humans, on the other hand, threaten the survival of bats in many ways. Some bats, such as the large flying foxes of Southeast Asia, are hunted for food. Species such as Mexican fishing bats have became locally extinct following the introduction by humans of cats and rats to their island habitats. Mining, waste disposal and irresponsible tourism threaten the caves where bats roost.

Bats have suffered grievously from the loss of their natural habitat to agriculture and forestry. A vicious cycle ensues. The decline of bats in agricultural areas causes farmers to use more chemicals to control insect pests, which risks poisoning the few remaining bats.

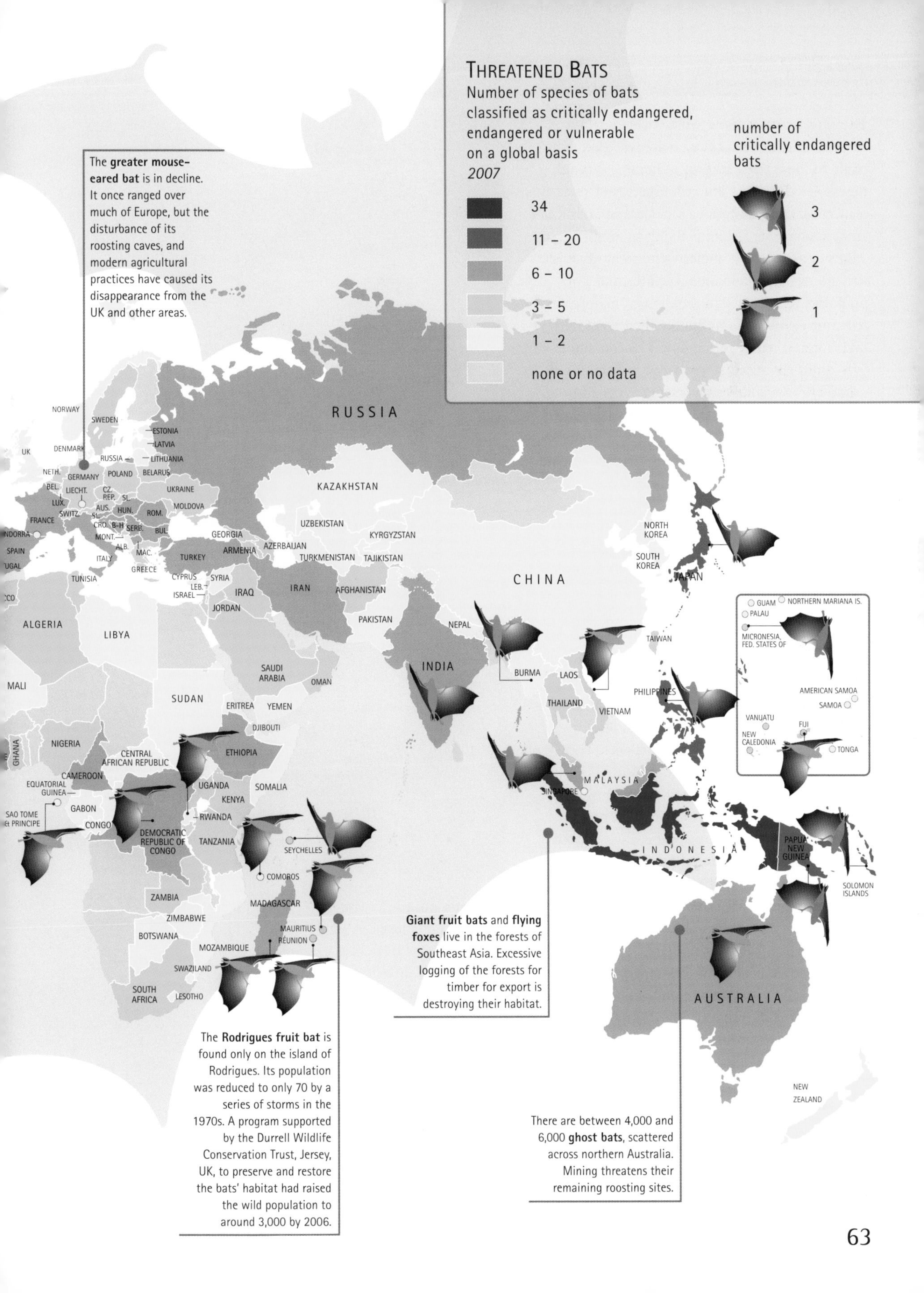

THREATENED BATS
Number of species of bats classified as critically endangered, endangered or vulnerable on a global basis
2007
34
11 – 20
6 – 10
3 – 5
1 – 2
none or no data
number of critically endangered bats
3
2
1
The **greater mouse-eared bat** is in decline. It once ranged over much of Europe, but the disturbance of its roosting caves, and modern agricultural practices have caused its disappearance from the UK and other areas.
RUSSIA
KAZAKHSTAN
CHINA
INDIA
ALGERIA
LIBYA
SUDAN
AUSTRALIA
INDONESIA
MALAYSIA
NORWAY
SWEDEN
ESTONIA
LATVIA
LITHUANIA
DENMARK
UK
NETH.
GERMANY
POLAND
BELARUS
BEL.
LIECHT.
CZ. REP.
SL.
UKRAINE
LUX.
AUS.
HUN.
MOLDOVA
SWITZ.
ROM.
FRANCE
CRO.
B-H
SERB.
BUL.
MONT.
ALB.
MAC.
ITALY
GREECE
SPAIN
GEORGIA
ARMENIA
AZERBAIJAN
TURKEY
UZBEKISTAN
KYRGYZSTAN
TURKMENISTAN
TAJIKISTAN
NORTH KOREA
SOUTH KOREA
JAPAN
TUNISIA
CYPRUS
SYRIA
LEB.
ISRAEL
IRAQ
IRAN
AFGHANISTAN
JORDAN
PAKISTAN
NEPAL
TAIWAN
BURMA
LAOS
SAUDI ARABIA
OMAN
MALI
THAILAND
VIETNAM
PHILIPPINES
ERITREA
YEMEN
DJIBOUTI
GHANA
NIGERIA
ETHIOPIA
CENTRAL AFRICAN REPUBLIC
CAMEROON
EQUATORIAL GUINEA
UGANDA
SOMALIA
KENYA
SINGAPORE
SAO TOME & PRINCIPE
GABON
CONGO
RWANDA
DEMOCRATIC REPUBLIC OF CONGO
TANZANIA
SEYCHELLES
COMOROS
ZAMBIA
MADAGASCAR
ZIMBABWE
MAURITIUS
RÉUNION
BOTSWANA
MOZAMBIQUE
SWAZILAND
SOUTH AFRICA
LESOTHO
PAPUA NEW GUINEA
SOLOMON ISLANDS
NEW ZEALAND
GUAM
NORTHERN MARIANA IS.
PALAU
MICRONESIA, FED. STATES OF
AMERICAN SAMOA
SAMOA
VANUATU
FIJI
NEW CALEDONIA
TONGA
Giant fruit bats and **flying foxes** live in the forests of Southeast Asia. Excessive logging of the forests for timber for export is destroying their habitat.
The **Rodrigues fruit bat** is found only on the island of Rodrigues. Its population was reduced to only 70 by a series of storms in the 1970s. A program supported by the Durrell Wildlife Conservation Trust, Jersey, UK, to preserve and restore the bats' habitat had raised the wild population to around 3,000 by 2006.
There are between 4,000 and 6,000 **ghost bats**, scattered across northern Australia. Mining threatens their remaining roosting sites.

Dolphins and Whales

Dolphins, porpoises and whales are known as cetaceans. They are aquatic mammals that must come to the water's surface to breathe through blowholes. There are about 80 species of cetaceans. Dolphin species include pilot and killer whales, and river dolphins living in South America and Asia. Porpoises tend to be smaller and stouter than dolphins. Dolphins and porpoises usually live in groups and mainly eat fish and squid. They are highly intelligent, and use sounds and ultrasonic pulses to communicate.

Dolphins that become ensnared in fishing nets usually drown. Those that live in rivers and estuaries are also adversely affected by chemical pollution and silting, and having to compete for food with humans. Their limited distribution makes them particularly vulnerable.

Some whales (including the sperm, bottle-nosed, beaked and beluga) have teeth and eat fish. Others (including the gray, right, humpback and blue) strain plankton and krill from the water, using a dense fringe of blade-shaped horny plates ("baleen") in their mouths. Some whales are social, traveling in groups. They use a range of underwater sounds: barks, whistles, screams, and moans, for communication, and high-intensity clicks for navigation and to identify food sources.

Dead whales yield several products of commercial value, including meat, oil, whalebone, and ambergris which is used in perfume. Hundreds of years of whaling by increasingly effective methods has led to the near-extinction of several species, in particular the blue whale. The International Whaling Commission, set up in 1946, has attempted to combat this threat – both to whales and to the livelihood of the whalers. A moratorium on whaling, which came into force in 1986, has reduced the number of whales killed, and some species, such as gray and humpback whales, are showing signs of recovery. The Japanese have encouraged countries to join the Commission and vote with them against the moratorium but, as of early 2008, without success.

For some communities, whaling provides an important part of their diet, and plays a key role in their economy and culture. These are exempt from the ban on commercial whaling. People from Greenland hunt fin and minke whales, those in Siberia hunt gray whales. In Alaska bowhead, and occasionally gray, whales are caught. Japanese fleets have also been permitted to kill hundreds of minke whales, as well as some fin whales and 50 humpback whales each year for scientific purposes.

Whales are affected by human development, as their breeding and calving areas are often in shallow coastal waters. But tourism, in the form of "whale watching", now helps support the economies of towns and villages that once relied on income from hunting whales. It also helps to promote public understanding and appreciation of whales, which may influence governments' inclination to continue to protect them.

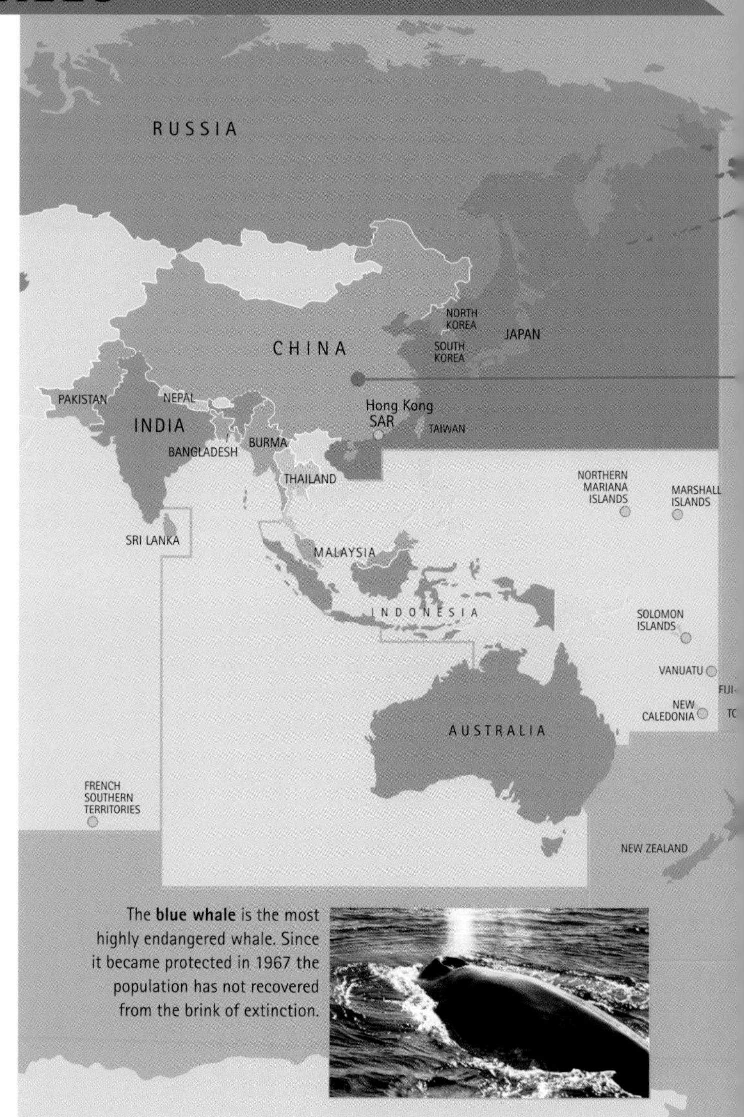

The **blue whale** is the most highly endangered whale. Since it became protected in 1967 the population has not recovered from the brink of extinction.

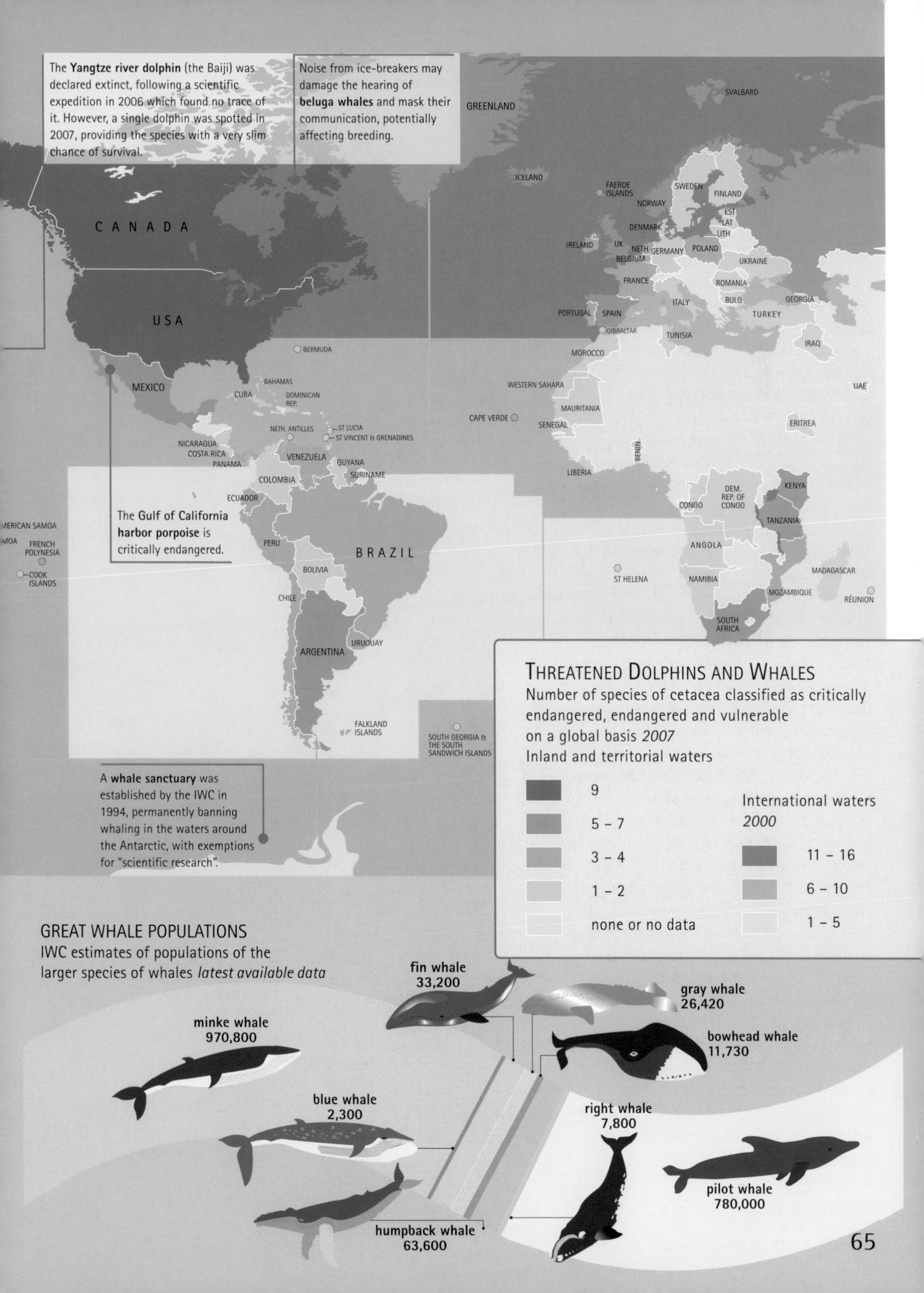

THREATENED DOLPHINS AND WHALES

Number of species of cetacea classified as critically endangered, endangered and vulnerable on a global basis *2007*

Inland and territorial waters

- 9
- 5 – 7
- 3 – 4
- 1 – 2
- none or no data

International waters *2000*

- 11 – 16
- 6 – 10
- 1 – 5

GREAT WHALE POPULATIONS

IWC estimates of populations of the larger species of whales *latest available data*

Reptiles and Amphibians

Reptiles are air-breathing vertebrates that have internal fertilization and scaly bodies. They include snakes, lizards, turtles, crocodiles and alligators.

The skins of lizards, crocodiles and snakes are used to make leather goods such as luggage, handbags and shoes. This has led to the virtual extinction of several species of crocodile and to a severe reduction in the populations of large lizards, snakes, and turtles.

Vertebrates such as frogs, toads and salamanders that are able to exploit both water and land habitats are known as amphibians (from the Greek for "living a double life"). Despite this description, however, some species are permanent land dwellers, while others are completely aquatic. Amphibians absorb oxygen through their skin, which is kept moist by mucus-secreting glands.

The Global Amphibian Assessment, published in 2005, indicated that almost a third of amphibian species were threatened with extinction. Because of their absorbent skins, amphibians are susceptible to water pollution, even at low concentrations, or in protected areas some distance from the source of pollution. They are therefore considered one of the best indicators of pollution in the environment.

Because amphibians rely on moisture, they are also affected by changes in weather patterns. In the Americas, the Caribbean and Australia frog numbers have plummeted in recent years, largely because of the fungus chytrid, which attacks keratin in their skin, producing toxins and preventing them from using their skins to breathe. Scientists now think that it may be linked to drought, which is becoming increasingly frequent because of recent climate change.

Since the early 1990s environmentally conscious Australians have been reintroducing tadpoles into ponds and streams in an attempt to bolster flagging populations. This not only spreads disease, but introduces alien genes or even alien species that displace native frogs. Exotic frog species are also being introduced accidentally, imported in crates of fruit, vegetables, and flowers.

The last wild **Panamanian golden frogs** were taken into captivity in 2006 to protect them from the chytrid fungus that attacks the skin of amphibians, preventing them from taking in oxygen. It is a major threat to amphibians in many parts of the world.

The **marine iguana** is only found on the Galapagos Islands and is therefore vulnerable to local pollution and the destruction of its habitat.

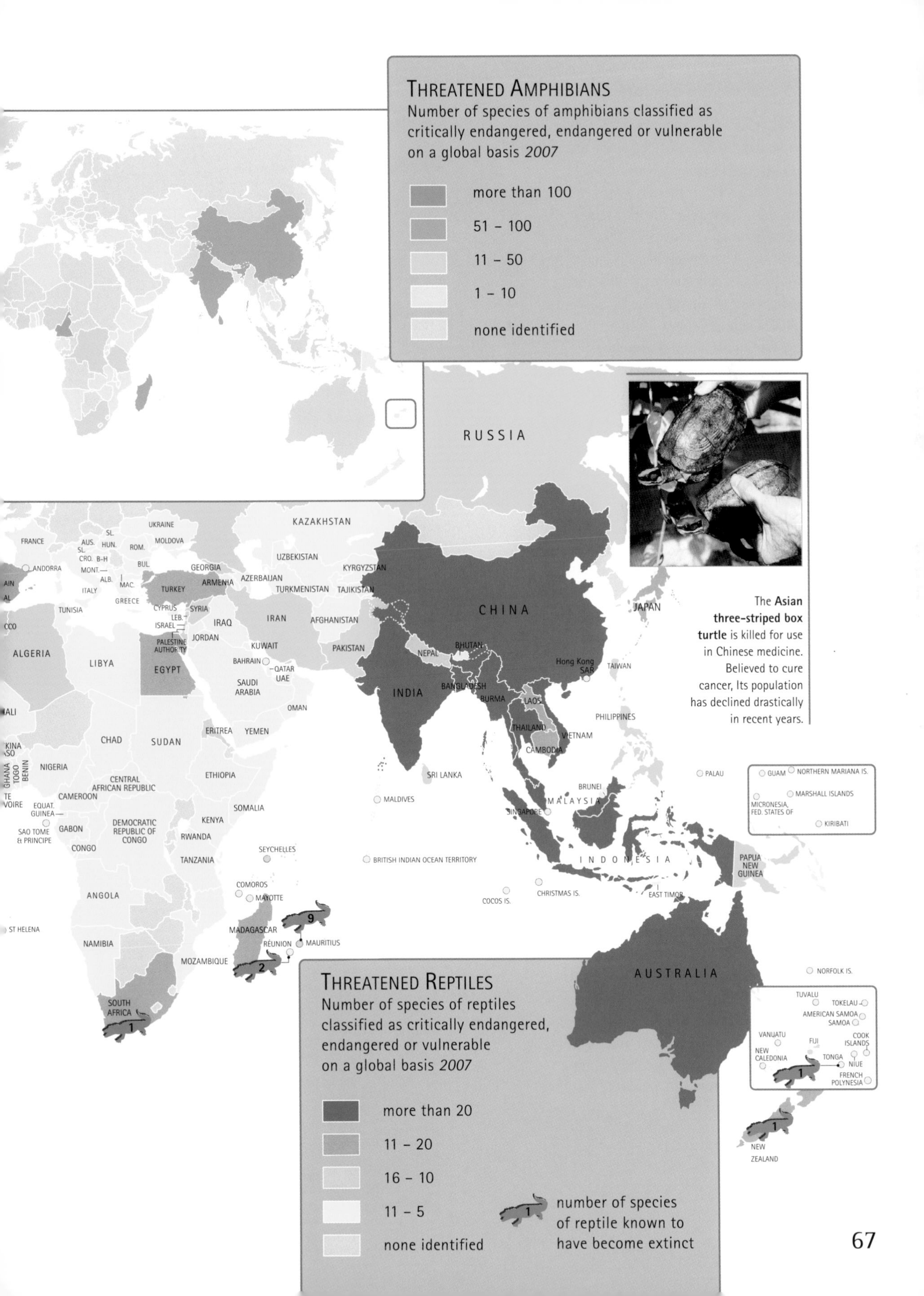

The **Asian three-striped box turtle** is killed for use in Chinese medicine. Believed to cure cancer, Its population has declined drastically in recent years.

INVERTEBRATES

Invertebrates are animals that lack backbones. They comprise over 90 percent of animal species. Some invertebrates are soft-bodied but many have a hard outer skeleton that provides protection and anchorage for muscles; these animals are called "arthropods". The three groups of arthropods are chelicerates (spiders, scorpions and mites), crustaceans (crabs and shrimps) and uniramians (insects and centipedes).

The range of size and habitat of invertebrates is staggering. Some plankton measure less than two-hundredths of an inch (0.5 mm), while giant squid can grow up to 33 feet (10 metres) long. Crustaceans have been found over 2.5 miles (4 km) under the sea and spiders have been seen on Mount Everest. The practical difficulties in catching and inspecting smaller invertebrates means that the number of species can only be estimated. Some experts suggest that the number of species of insects alone is over 10 million and more are being discovered every year in remote areas of rainforest.

As new invertebrates are identified, other species are being added to the IUCN Red List. Identifying a species under threat requires money and effort. This map, together with others in the atlas, reveals a strong correlation between the number of species of invertebrates recognized as under threat in a country and the investment put into investigating them.

Invertebrates play a crucial role in many ecosystems. Other animals often rely on them for food. The plankton of the oceans support higher forms of life, including whales. Invertebrates can help in other ways too – insects pollinate flowering plants, and worms mix and aerate soil.

Some invertebrates, such as butterflies, capture the public imagination, but most are perceived of as distasteful or at least too numerous to warrant concern for their conservation. Invertebrates with restricted habitats or limited mobility are at risk both from human development and from climate change, which could, for example, eliminate a species that occurs only on a single mountain.

Monarch butterflies migrate to Mexico for the winter from as far away as Canada. Almost all over-wintering Monarchs are concentrated in just eight groves of trees, with as many as 100,000 huddling together on a single tree for warmth. Disturbance by loggers can cause Monarchs to scatter on arrival in the autumn, jeopardizing their ability to mate the following spring. Climate change will bring more rain to these groves and the ice that forms kills Monarchs.

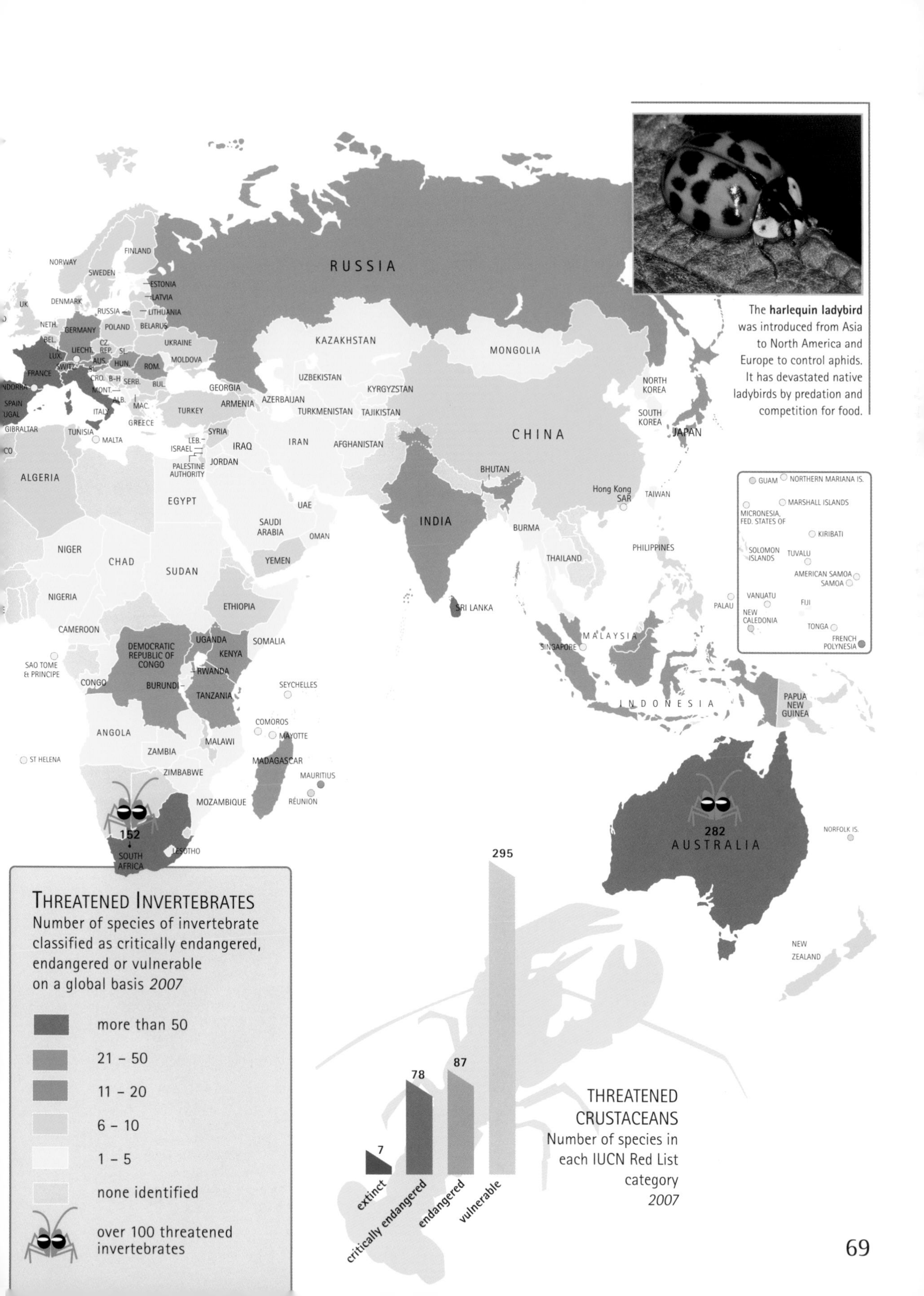
The harlequin ladybird was introduced from Asia to North America and Europe to control aphids. It has devastated native ladybirds by predation and competition for food.
RUSSIA
NORWAY
SWEDEN
FINLAND
ESTONIA
LATVIA
LITHUANIA
DENMARK
UK
RUSSIA
BELARUS
NETH.
BEL.
GERMANY
POLAND
LUX.
CZ. REP.
SL.
UKRAINE
LIECHT.
SWITZ.
AUS.
HUN.
MOLDOVA
FRANCE
SI.
ROM.
CRO.
B-H
SERB.
BUL.
MONT.
ALB.
MAC.
ITALY
GREECE
SPAIN
GIBRALTAR
TUNISIA
MALTA
ALGERIA
GEORGIA
ARMENIA
AZERBAIJAN
TURKEY
SYRIA
LEB.
ISRAEL
PALESTINE AUTHORITY
JORDAN
IRAQ
IRAN
EGYPT
KAZAKHSTAN
UZBEKISTAN
KYRGYZSTAN
TURKMENISTAN
TAJIKISTAN
AFGHANISTAN
MONGOLIA
NORTH KOREA
SOUTH KOREA
JAPAN
CHINA
BHUTAN
Hong Kong SAR
TAIWAN
INDIA
BURMA
THAILAND
PHILIPPINES
UAE
SAUDI ARABIA
OMAN
YEMEN
NIGER
CHAD
SUDAN
NIGERIA
ETHIOPIA
SRI LANKA
CAMEROON
SOMALIA
UGANDA
KENYA
DEMOCRATIC REPUBLIC OF CONGO
RWANDA
BURUNDI
TANZANIA
SAO TOME & PRINCIPE
CONGO
SEYCHELLES
COMOROS
MAYOTTE
ANGOLA
MALAWI
ZAMBIA
MADAGASCAR
ST HELENA
ZIMBABWE
MAURITIUS
MOZAMBIQUE
RÉUNION
152
SOUTH AFRICA
LESOTHO
MALAYSIA
SINGAPORE
INDONESIA
PALAU
PAPUA NEW GUINEA
GUAM
NORTHERN MARIANA IS.
MARSHALL ISLANDS
MICRONESIA, FED. STATES OF
KIRIBATI
SOLOMON ISLANDS
TUVALU
AMERICAN SAMOA
SAMOA
VANUATU
FIJI
NEW CALEDONIA
TONGA
FRENCH POLYNESIA
282
AUSTRALIA
NORFOLK IS.
NEW ZEALAND
THREATENED INVERTEBRATES
Number of species of invertebrate classified as critically endangered, endangered or vulnerable on a global basis 2007
more than 50
21 – 50
11 – 20
6 – 10
1 – 5
none identified
over 100 threatened invertebrates
THREATENED CRUSTACEANS
Number of species in each IUCN Red List category 2007
7
78
87
295
extinct
critically endangered
endangered
vulnerable

FISH

Fish are found in rivers, lakes and oceans all over the world, and range in size from a tiny goby, half an inch (12.5 mm) in length, to the 50-foot (15-metre) whale shark. Some species range widely through the oceans, while others are restricted to a single lake.

Fish congregate in schools to feed, breed and minimize the risk (to each individual) from predators. They lay eggs that hatch into larvae and develop into adult fish. Larvae and adult fish have different needs, and often live far apart. Oceanic fish may, for example, have coastal or freshwater spawning grounds.

Over 90 percent of the global catch comprises saltwater fish. Countries have control over fisheries in an "exclusive economic zone" up to 200 miles (320 km) from their coasts – or to the midpoint between opposing coasts. Despite this, and measures to define fishing quotas, most commercial fisheries are seriously depleted. Fish that move between national fishing zones are hard to protect. Stocks of the critically endangered southern bluefin tuna are only 10 percent of their level in the 1950s, but although a commission was established in 1994 to conserve the species, disputes between its member nations over quotas have hampered progress.

Freshwater fish, which represent only 10 percent of all fish caught, tend to be eaten locally, often providing crucial dietary protein for rural societies. Because freshwater species occur in limited areas, they are especially vulnerable to over-fishing and ecological damage. Dams and channels fragment habitats. Effluent and agricultural chemicals pollute them. Water abstraction sucks rivers dry.

Aquaculture (fish farming) accounts for about one-third of all fish produced. Around 90 percent comes from Asia, but the industry is growing rapidly in Latin America and Africa. Although it is hailed by some as a solution to the depletion of wild fish stocks, the production of a given weight of carnivorous fish such as salmon or of shrimp requires at least twice that in protein, much of it derived from captured wild fish. The waste emitted by unregulated enterprises can be highly damaging to fragile coastal ecosystems such as mangroves. Human health may also suffer from contaminants passed down in fish feed.

In 1997, WWF and Unilever founded the Marine Stewardship Council (MSC), which established criteria for good fishing management. In Alaska, where overfishing reduced salmon harvests to about 25 million fish in 1959, habitat protection allowed stocks to recover sufficiently to enable a commercial salmon catch of 168 million fish in 2004.

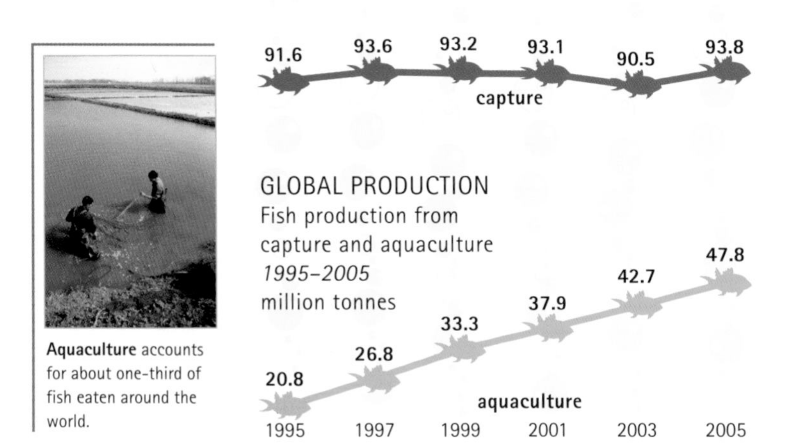

Aquaculture accounts for about one-third of fish eaten around the world.

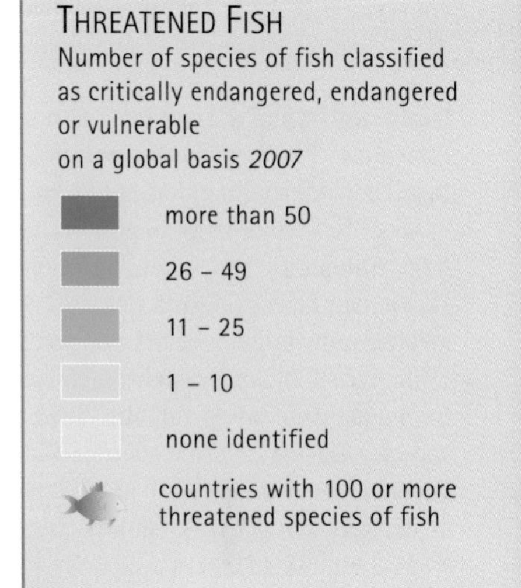

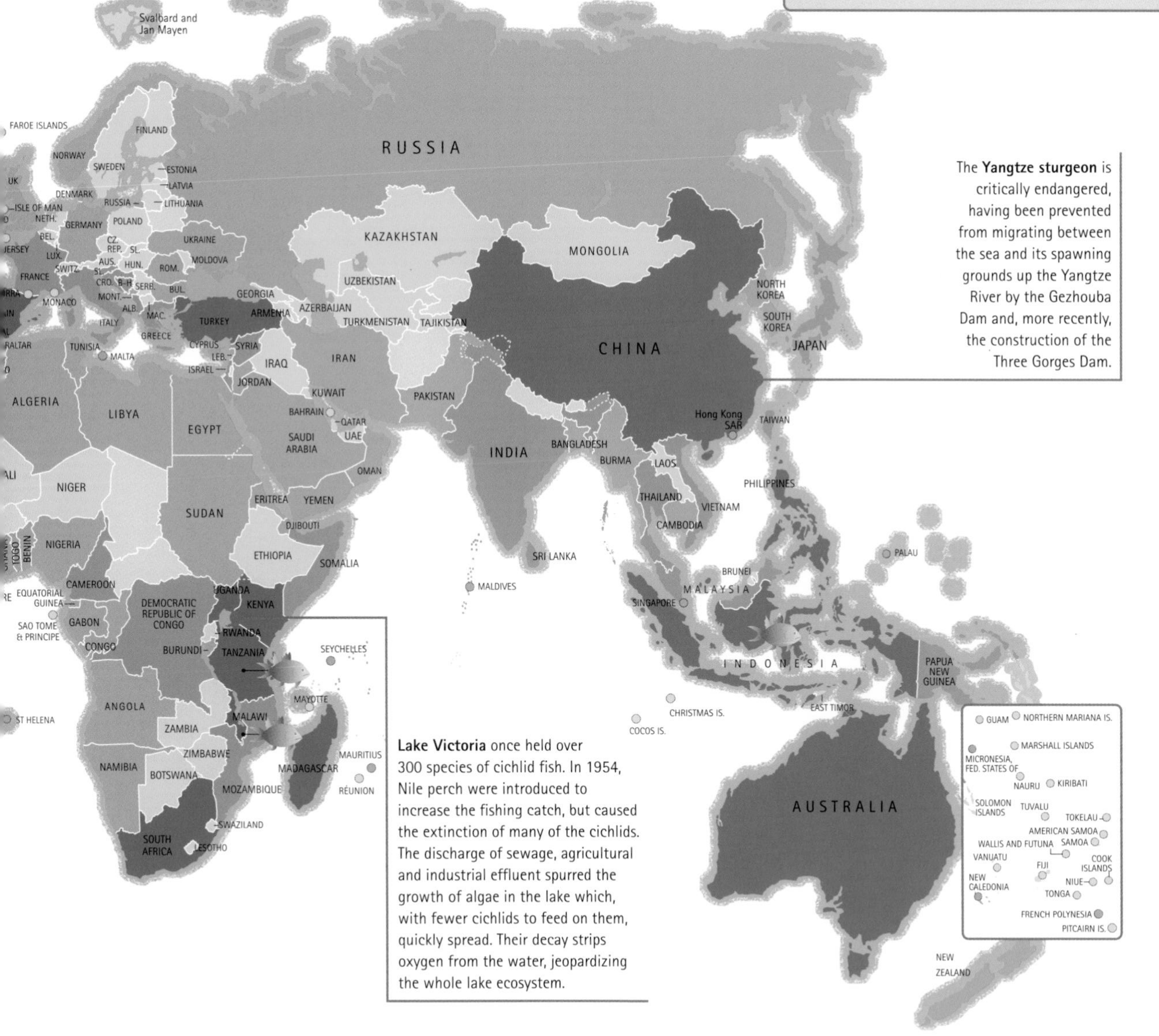

The **Yangtze sturgeon** is critically endangered, having been prevented from migrating between the sea and its spawning grounds up the Yangtze River by the Gezhouba Dam and, more recently, the construction of the Three Gorges Dam.

Lake Victoria once held over 300 species of cichlid fish. In 1954, Nile perch were introduced to increase the fishing catch, but caused the extinction of many of the cichlids. The discharge of sewage, agricultural and industrial effluent spurred the growth of algae in the lake which, with fewer cichlids to feed on them, quickly spread. Their decay strips oxygen from the water, jeopardizing the whole lake ecosystem.

PLANTS

Plants are vital to the survival of all other life forms on earth, forming the basis of most animal food chains, including that of humans. Plants harness the sun's energy to produce their own food through the process of photosynthesis. They absorb carbon dioxide and release oxygen, helping to maintain a healthy atmosphere.

The plant kingdom is incredibly diverse, ranging from simple mosses to complex flowering plants and the 300-foot (100-meter) high redwood tree. There are between 275,000 and 300,000 species of vascular plants (ferns, conifers and flowering plants), around 34,000 of which are threatened with extinction (see right).

Humans often carry plants and their seeds around the world, sometimes intentionally to grow for food and for decoration, sometimes inadvertently. Predators and diseases that afflict native plants are often not well adapted to attack these introduced, "exotic", plant species. As a result some exotic species become "invasive", competing with and displacing native species of plants. Exotic species may also interbreed with native plants, giving rise to hybrids. Plants endemic to a small area of land or an island are particularly at risk. Native plants on islands such as Hawaii and St. Helena, for example, have been devastated by invasive species.

Habitat destruction is an even greater threat to plant diversity. Human settlement and intensive farming continue to encroach on natural areas, threatening plants such as the Alabama canebrake pitcher-plant, which is now critically endangered following drainage of over 50 percent of the wetlands of Alabama. Unsustainable logging in forests around the world also threatens the survival of many endemic populations of plants (see pages 22-25), just as the world is waking up to the vital medicinal qualities of many species.

The **lady's slipper orchid** grows on limestone soils across the northern hemisphere. Because of its beauty it is prized by plant collectors, and has therefore always been rare. The remaining plants suffer from habitat destruction by logging and agriculture.

EXTENT OF THE THREAT
Number of species of plant in each IUCN Red List category
2007

Threatened Plants

Number of species of plant classified as critically endangered, endangered or vulnerable on a global basis *2007*

- more than 200
- 101 – 300
- 51 – 100
- 11 – 50
- 1 – 10
- none identified
- more than 300 threatened plants, number given

Trampling, destruction of woodland, trade in wild bulbs and hybridization with its domestic relatives has caused the decline of the **wild bluebell** in the UK, which hosts between a quarter and a half of the global population.

Over 1,000 species of plants grow in the alpine regions of the **Caucasus**. In these rugged mountains plant populations became isolated and have evolved into new species that are now threatened by over-grazing.

The **mandrinette** is a species of hibiscus that produces a scarlet flower. Only 36 mature trees survive, all of them on Mauritius. Competition from invasive plants has prevented mandrinettes from reproducing.

The **bastard quiver tree** is found in South Africa and Namibia. Over-grazing by goats and donkeys has reduced the population to fewer than 3,000.

ENDANGERED BIRDS

5

"The sedge is wither'd from the lake,
And no birds sing."

– John Keats,
"La Belle Dame Sans Merci" (1817)

BIRDS

Birds are warm-blooded vertebrates that have developed hollow bones, wings and feathers that enable most, but not all, to fly. Some birds are generalists, eating a variety of seeds, nectar, fruit, invertebrates or other animals. Other species have specialized diets. In 2008, an estimated 12 percent (over 1,200) of all known species of bird were considered under threat.

The greatest threat to birds is human destruction of their habitat through logging, commercial development and intensive agriculture, and the introduction of new predators, competitors and diseases. Even domestic cats, if allowed to roam free, can kill hundreds of small birds each year. Uncontrolled hunting and trapping threatens many species of birds. The imperial woodpecker, which once lived in upland forest in north eastern Mexico, was one of many species driven to extinction in the 20th century by deforestation and hunting. Migratory birds are especially vulnerable to trapping and shooting.

Intensive agriculture also threatens farmland birds, especially in North America and Europe. Winter sowing produces crops too dense for birds to nest in, and the frequent cutting of summer grass for silage destroys their nest sites. Skylark populations have decreased sharply in western Europe, but in eastern Europe, where intensive methods are not yet widespread, the bird's populations tend to be more stable.

Climate change worldwide affects the fragile balance between birds and their food supply. Plants are flowering earlier in the spring, and changes to vegetation are altering the abundance of insects, worms and other creatures. This is likely to have a significant impact on the breeding and migratory patterns of birds.

In Southeast Asia, Africa and South America deforestation is occurring at a rapid rate (see pages 42-43), and at each review more tropical rainforest birds are classified as threatened. Rainforest species, such as the Indonesian caerulean paradise-flycatcher, are most at risk from unsustainable logging and forest clearance for agriculture and exotic timber plantations.

In 1998 the **bald eagle** was officially removed from the US Fish and Wildlife Service critical list. Extensive use of pesticides had reduced its population in the USA (excluding Alaska) to fewer than 500, but conservation measures resulted in more than 7,000 nesting pairs by 2008.

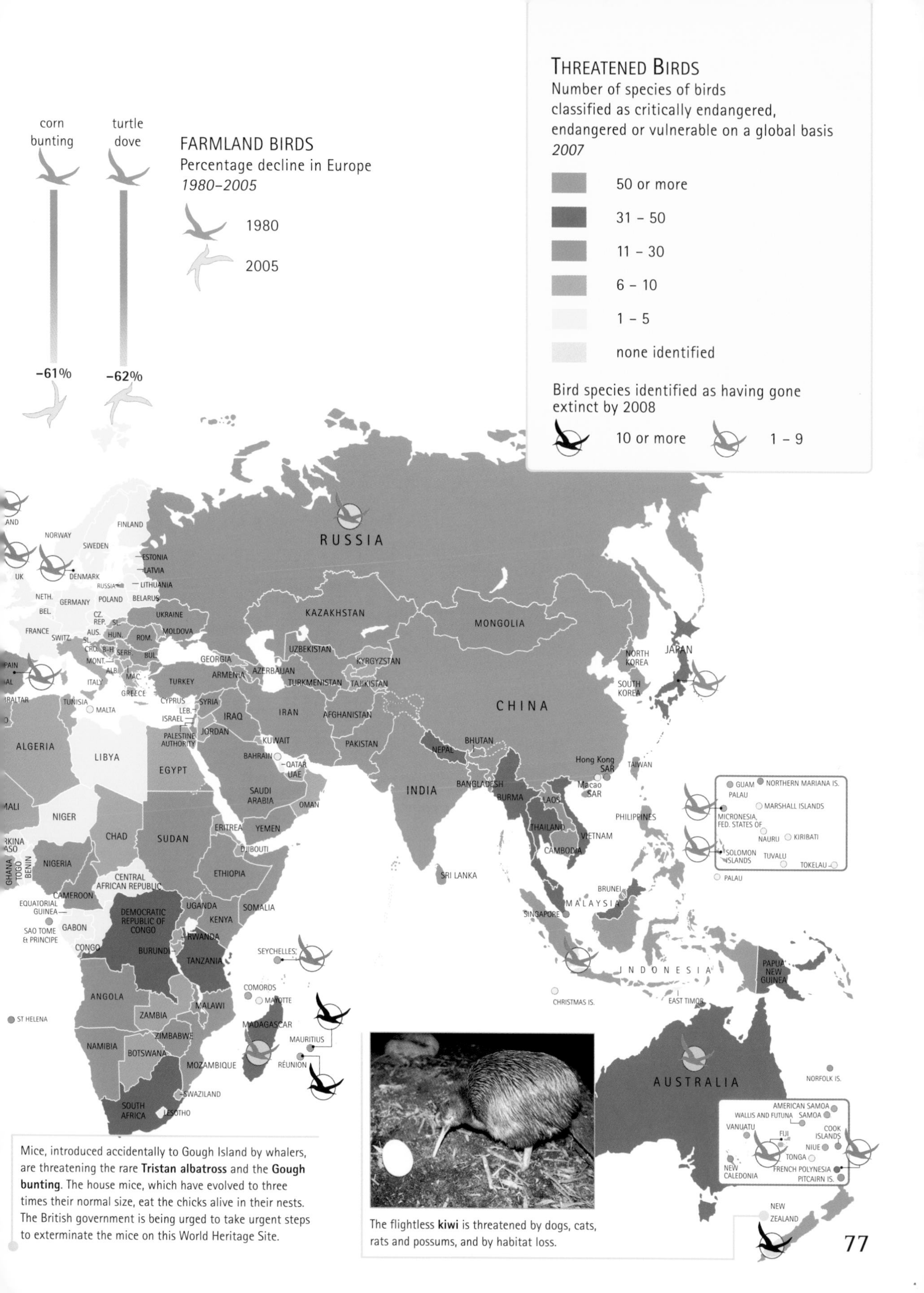

Mice, introduced accidentally to Gough Island by whalers, are threatening the rare **Tristan albatross** and the **Gough bunting**. The house mice, which have evolved to three times their normal size, eat the chicks alive in their nests. The British government is being urged to take urgent steps to exterminate the mice on this World Heritage Site.

The flightless **kiwi** is threatened by dogs, cats, rats and possums, and by habitat loss.

Birds of Prey

Birds of prey include owls, and raptors such as hawks, eagles, vultures, and falcons. All birds of prey have talons for seizing their quarry, and curved beaks for tearing its flesh. They have good vision and hearing, but often a poor sense of smell. Raptors hunt during the day, and owls are nocturnal. Raptors eat other birds, mammals and carrion. Some, such as the bald eagle, can pluck fish from water. They are all skillful fliers and larger species have long, broad wings for soaring.

Many birds of prey have suffered from loss of habitat. Each breeding pair requires a large territory, which is often difficult to protect from development. Many of the chemicals used in farming are toxic to birds of prey. For example, the organochlorine pesticide DDT causes the shells of eagles' eggs to thin, making them prone to break. Prohibition of the use of such pesticides in developed countries has helped protect birds of prey, but developing countries often cannot afford the less damaging but more expensive alternatives. Birds such as the Steller's sea eagle, which feed on animals that have been shot, often suffer from lead poisoning as a result. Other birds of prey are killed for sport and food, and captured for falconry – all of which further depletes the populations of some species.

Habitat protection plays a critical role in the conservation of birds of prey. Control of poaching is important too. In many countries it is illegal to hunt birds of prey, steal their eggs or disturb their nests. Captive breeding and rehabilitation centers for injured birds can significantly help species at risk of extinction.

The **California condor** population fell below 100 in 1940 because of hunting and lead poisoning. In 1987 the remaining nine birds were captured for a breeding program at Los Angeles Zoo and at the San Diego Wild Animal Park. This has held the species back from the brink of extinction and by 2008 there were 146 Californian condors living in the wild, and 151 in captivity.

CALIFORNIAN CONDOR IN CRISIS
Number of individual birds in the wild

THREATENED BIRDS OF PREY

Number of species of birds of prey classified as critically endangered, endangered or vulnerable on a global basis *2007*

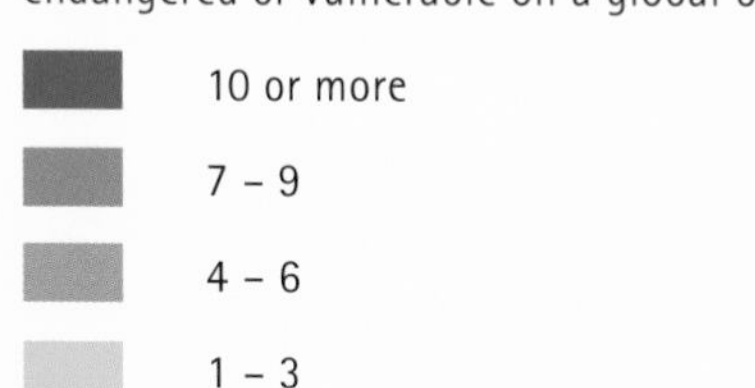

The **great Philippine eagle**, which produces a single chick every two years, is vulnerable to clearance of its rainforest habitat for logging and farming. No more than 300 survive, almost all on the island of Mindanao.

DECLINE OF ASIAN WHITE-BACKED VULTURE

Breeding population in Keoladeo National Park, near Bharatpur, India

The **white-backed** and **long-billed vultures,** once seen circling in huge flocks in the Indian subcontinent, are now nearing extinction, and attempts to breed from captive vultures have met with little success. The vultures have been killed off by consuming the carcasses of cattle that had been given the drug diclofenac to treat arthritis and boost milk yields. This causes kidney failure in the vultures and leads to death within 30 days. The birds performed the important function of consuming the dead bodies of livestock and, for those of the Parsee faith, of humans. This role is increasingly being filled by feral dogs, leading to an alarming increase in rabies.

Parrots and Cockatoos

There are over 300 species of parrot, including lorikeets, cockatoos and parakeets. As many as 95 of them, or 28 percent, are threatened – the highest proportion of any of the major bird families.

Parrots come in a variety of sizes and shapes. The pygmy parrot can be less than 3 inches (10 cm) long, while some macaws grow to over 3 feet (1 metre) in length. Most are brightly coloured. Parrots have strong, hooked beaks, which they use for breaking open nuts. Many plants, such as the oil palm, rely on parrots to disperse their seeds. A few parrots are predatory. For example, the New Zealand kea is known to attack sheep.

Most parrots live in tropical forests, nesting in tree holes. They are not built for flying long distances, which means that species are often restricted to isolated islands, making them especially vulnerable. In the West Indies at least 16 species are known to have been exterminated by European explorers, and the Puerto Rican parrot is critically endangered (see right).

Parrots, particularly the African Gray and Amazon, are excellent mimics. This ability, along with their friendly nature and lively plumage, make them popular in zoos and also as pets, with most demand coming from Europe and Japan. Some countries, such as Mexico and Australia, forbid the export of parrots in order to conserve wild populations. Despite such efforts, the legal and illegal trade in parrots is huge, threatening the viability of wild populations. Poaching alone affects 39 species and is a greater cause of mortality than natural causes.

Parrots are hunted for food, and some seed-eating parrots are hunted by farmers because they are considered to be pests. Introduced predators, such as cats, are a hazard and imported birds can displace native parrot species or introduce avian diseases.

The greatest threat to most parrot species is loss of habitat, from fire, logging and human settlement. The Species Survival Commission's conservation action plan for parrots presents techniques for their conservation in the face of habitat destruction and illegal hunting. Scarce resources in the face of competing economic interests may frustrate its effective adoption.

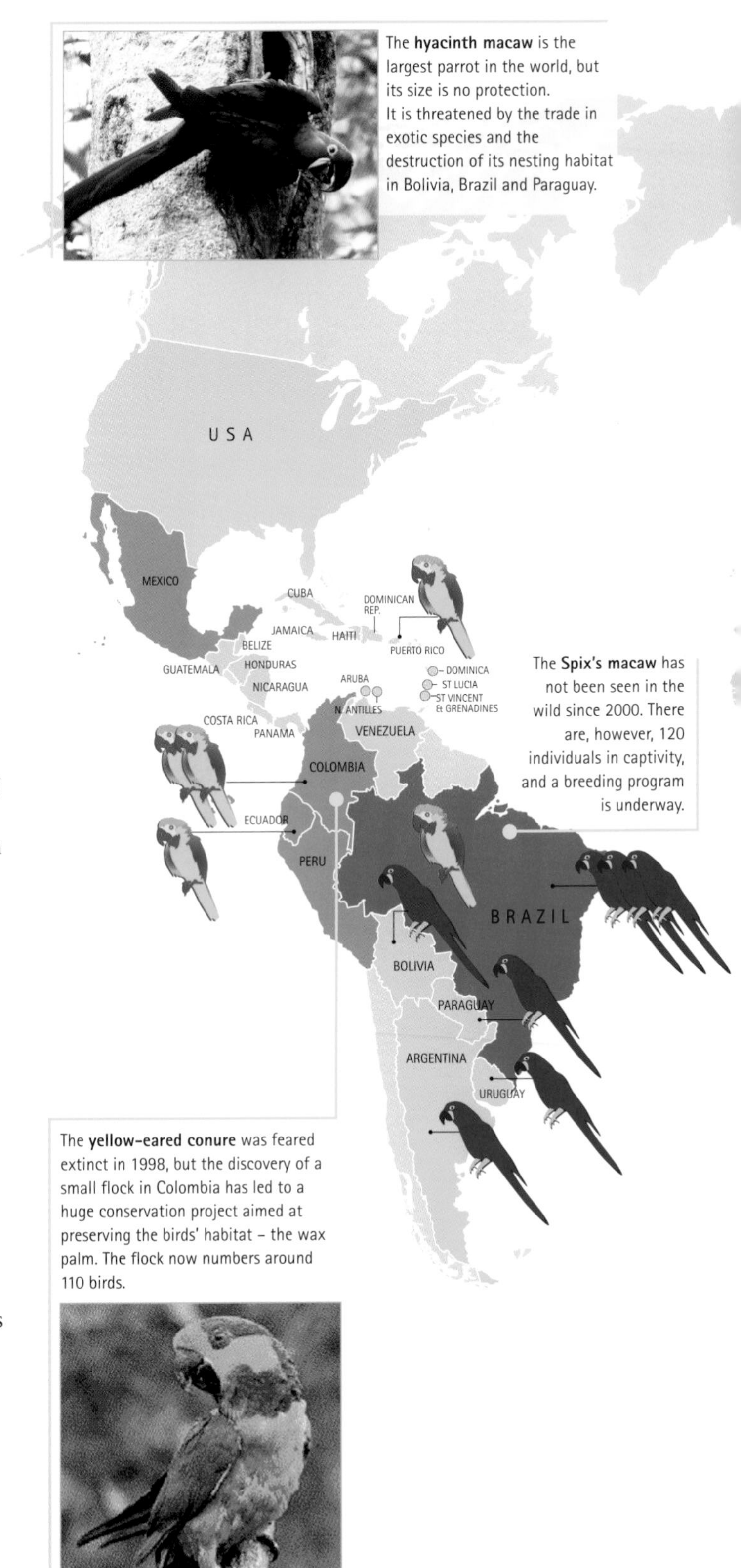

The **hyacinth macaw** is the largest parrot in the world, but its size is no protection. It is threatened by the trade in exotic species and the destruction of its nesting habitat in Bolivia, Brazil and Paraguay.

The **Spix's macaw** has not been seen in the wild since 2000. There are, however, 120 individuals in captivity, and a breeding program is underway.

The **yellow-eared conure** was feared extinct in 1998, but the discovery of a small flock in Colombia has led to a huge conservation project aimed at preserving the birds' habitat – the wax palm. The flock now numbers around 110 birds.

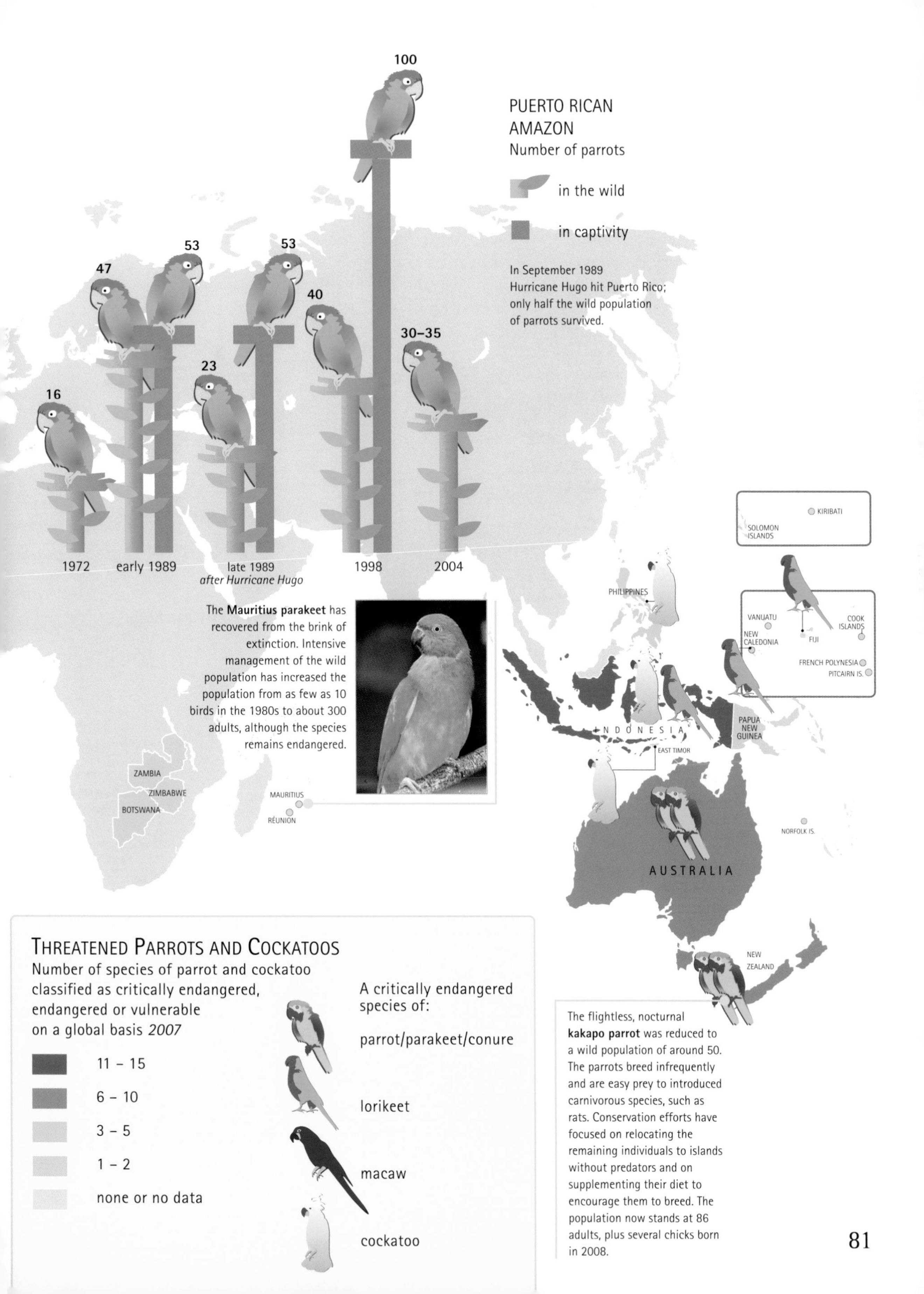

In September 1989 Hurricane Hugo hit Puerto Rico; only half the wild population of parrots survived.

The **Mauritius parakeet** has recovered from the brink of extinction. Intensive management of the wild population has increased the population from as few as 10 birds in the 1980s to about 300 adults, although the species remains endangered.

The flightless, nocturnal **kakapo parrot** was reduced to a wild population of around 50. The parrots breed infrequently and are easy prey to introduced carnivorous species, such as rats. Conservation efforts have focused on relocating the remaining individuals to islands without predators and on supplementing their diet to encourage them to breed. The population now stands at 86 adults, plus several chicks born in 2008.

SEABIRDS

The term "seabird" is not a taxonomic classification, but simply a description of birds that live in a certain habitat, and exhibit certain physical characteristics and patterns of behavior. Many seabird species are, for example, good fliers, often exploiting prevailing winds to migrate extraordinary distances. Penguins and auks, though flightless, are strong and agile swimmers. Most seabirds catch fish and crustaceans near the ocean surface, but cormorants can dive up to 150 feet (46 metres) below the surface in search of prey. Some seabirds forage on land for insects, small rodents and rubbish. Skuas and frigatebirds harass smaller seabirds returning to land with fish, forcing their victim to drop its catch so that they can pluck it from the air before it hits the sea.

Seabirds usually nest in huge colonies, sometimes of over a million birds, on islands or cliffs where they are reasonably safe from native predatory mammals – although not always from "introduced" species. Seabirds have a low rate of reproduction, which makes them vulnerable to catastrophic events, such as oil pollution or sudden food shortages.

In the 19th century North Atlantic sailors caused the extinction of the great auk by killing whole colonies for their flesh and feathers. Seabirds are still exploited for their eggs in the Faroe Islands, Iceland, Greenland and Southeast Asia, and such "harvesting", if not properly controlled, can threaten their survival. Far out at sea, birds are still not safe from humans. Drift-nets, nicknamed "walls of death", used by large fishing boats, trap birds such as shearwaters, despite a 1992 UN moratorium on the use of nets more than 1.4 miles (2.5 km) long.

Long-line fishing also takes a huge toll of seabirds. Lines can be up to 62 miles (100 km) long, with more than 20,000 hooks. The bait on the hooks presents a ready meal for seabirds – and also the risk of becoming caught and drowned. Conservation organizations are working with the fishing industry to promote responsible fishing techniques, such as an effective bird-scarer over the fishing line, and the weighting of the line to ensure that it sinks rapidly.

Pollution is an insidious threat to seabirds. Oil-spills have contaminated Adelie Penguins in the Antarctic, and when the *Exxon Valdez* spilled 11 million gallons of crude oil into Prince William Sound in Alaska, thousands of seabirds died. Global climate change threatens all seabirds, because plankton, which are fundamental to the marine food chain, become less abundant when seawater warms.

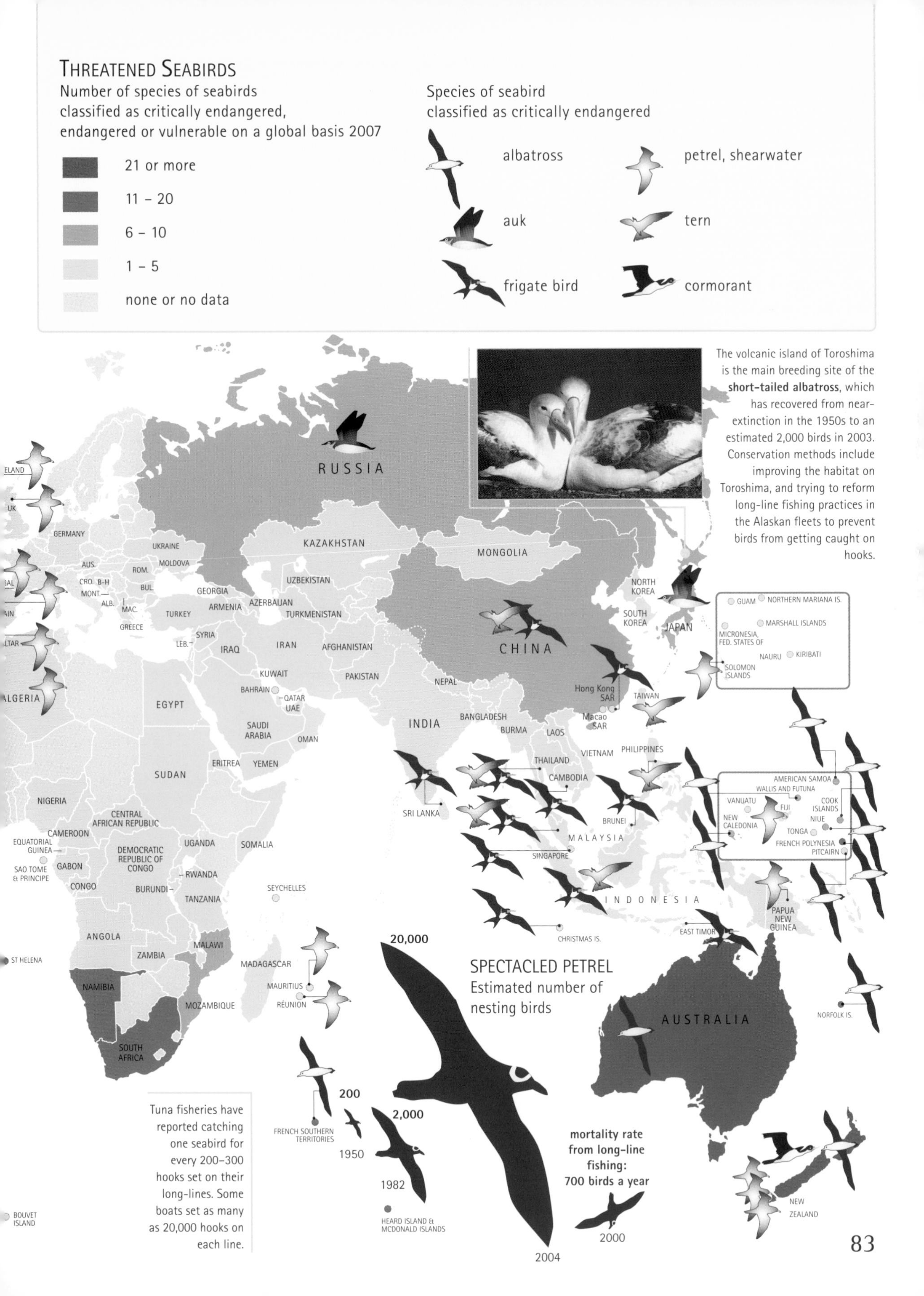
THREATENED SEABIRDS
Number of species of seabirds classified as critically endangered, endangered or vulnerable on a global basis 2007
21 or more
11 – 20
6 – 10
1 – 5
none or no data
Species of seabird classified as critically endangered
albatross
petrel, shearwater
auk
tern
frigate bird
cormorant
The volcanic island of Toroshima is the main breeding site of the **short-tailed albatross**, which has recovered from near-extinction in the 1950s to an estimated 2,000 birds in 2003. Conservation methods include improving the habitat on Toroshima, and trying to reform long-line fishing practices in the Alaskan fleets to prevent birds from getting caught on hooks.
RUSSIA
KAZAKHSTAN
MONGOLIA
CHINA
UKRAINE
MOLDOVA
GERMANY
AUS.
ROM.
CRO.
B.-H.
MONT.
BUL.
ALB.
MAC.
GREECE
TURKEY
GEORGIA
ARMENIA
AZERBAIJAN
UZBEKISTAN
TURKMENISTAN
SYRIA
LEB.
IRAQ
IRAN
AFGHANISTAN
PAKISTAN
KUWAIT
BAHRAIN
QATAR
UAE
SAUDI ARABIA
OMAN
YEMEN
ERITREA
EGYPT
SUDAN
NEPAL
INDIA
BANGLADESH
BURMA
LAOS
THAILAND
VIETNAM
CAMBODIA
PHILIPPINES
SRI LANKA
BRUNEI
MALAYSIA
SINGAPORE
INDONESIA
CHRISTMAS IS.
EAST TIMOR
PAPUA NEW GUINEA
NORTH KOREA
SOUTH KOREA
JAPAN
Hong Kong SAR
Macao SAR
TAIWAN
GUAM
NORTHERN MARIANA IS.
MARSHALL ISLANDS
MICRONESIA, FED. STATES OF
NAURU
KIRIBATI
SOLOMON ISLANDS
AMERICAN SAMOA
WALLIS AND FUTUNA
VANUATU
FIJI
COOK ISLANDS
NEW CALEDONIA
NIUE
TONGA
FRENCH POLYNESIA
PITCAIRN
NORFOLK IS.
AUSTRALIA
NEW ZEALAND
ELAND
UK
ALGERIA
NIGERIA
CAMEROON
CENTRAL AFRICAN REPUBLIC
EQUATORIAL GUINEA
SAO TOME & PRINCIPE
GABON
CONGO
DEMOCRATIC REPUBLIC OF CONGO
UGANDA
SOMALIA
RWANDA
BURUNDI
TANZANIA
SEYCHELLES
ANGOLA
ZAMBIA
MALAWI
ST HELENA
NAMIBIA
MOZAMBIQUE
MADAGASCAR
MAURITIUS
RÉUNION
SOUTH AFRICA
BOUVET ISLAND
FRENCH SOUTHERN TERRITORIES
HEARD ISLAND & MCDONALD ISLANDS
SPECTACLED PETREL
Estimated number of nesting birds
200
1950
2,000
1982
20,000
2004
mortality rate from long-line fishing: 700 birds a year
2000
Tuna fisheries have reported catching one seabird for every 200–300 hooks set on their long-lines. Some boats set as many as 20,000 hooks on each line.

Migratory Birds

Most migrating birds breed in the northern hemisphere in early summer, hatch their young and then head south to escape the winter. Their mobility makes them difficult to study and there is still much to be discovered about bird migration. Many projects are underway, involving ring tagging and satellite tracking. More traditional methods, relying on teams of amateur bird spotters, have been greatly helped by the ability to communicate information swiftly over the internet.

The main routes adopted by migrating birds have been identified and named, but not all birds stick to a single route. Some start on one flyway and then cross to another, and many birds take a circular tour – flying out on one route and back on another. Most migrating birds rely on stopover sites, at which they can rest, feed, and fortify themselves for the next stage of their journey. Wetlands are perhaps the most important of these habitats, but many, such as those near the Mediterranean coast, are threatened by commercial development, drainage for agriculture, or the loss of their inflow of fresh water.

Climate change will affect migratory birds in a number of ways. Unpredictable weather patterns and increased frequency and strength of storms will make it harder for small migrating birds to complete their journeys, putting these species under threat. Milder winters may increase the populations of non-migrating birds that compete for food with migrating ones, but may also mean that long-distance migrants do not need to travel so far to find temperate conditions. Drastic changes to rainfall patterns may mean the loss of vital habitats, such as the wetland areas on which so many migrants depend. In the Arctic, birds may lose their breeding grounds as permafrost melts and enables trees to colonize what was open tundra.

Many migrating birds never arrive at their destination. Millions are shot down, by farmers in the Middle East and Africa, and by hunters in southern Europe, where they are killed for sport. And if they escape the bullets, they still have to negotiate built-up areas, and avoid flying into skyscrapers or becoming entangled in power cables.

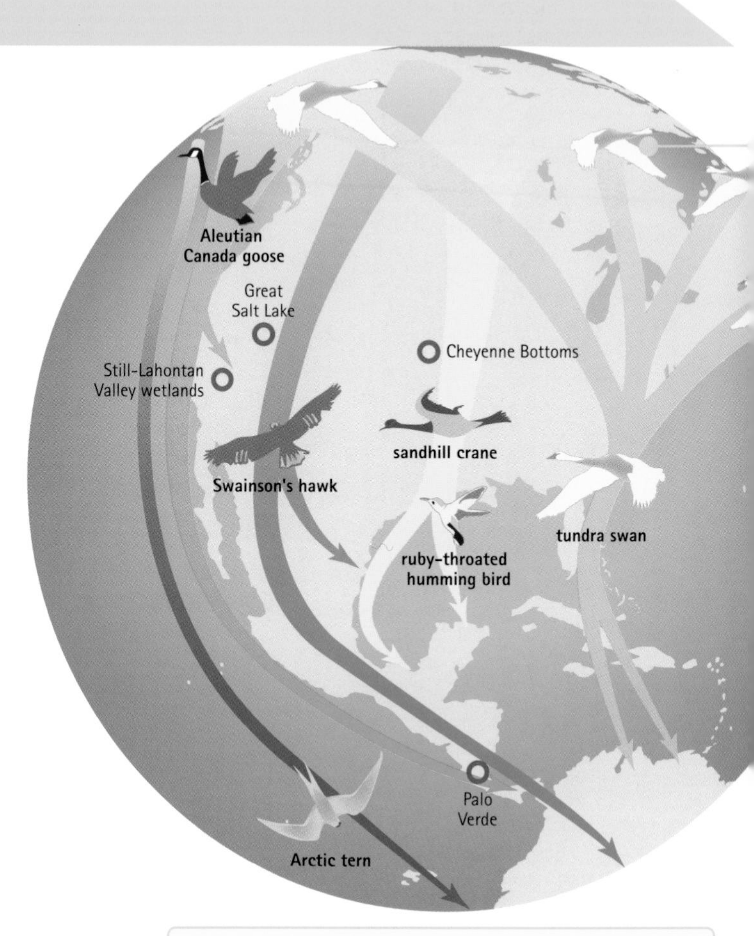

Migration Routes

Main routes taken by migrating birds to escape northern winter

Routes in Americas:
- Pacific Flyway
- Central Flyway
- Mississippi Flyway
- Atlantic Flyway
- route of the Arctic tern

Routes in Europe and Africa:
- Iceland to northern Europe
- Siberia to northern Europe
- East Atlantic Flyway
- Mediterranean/Black Sea Flyway
- Asia–Africa Flyway
- route of the Arctic tern

Routes in Asia and Oceania:
- Central Asian–South Asian Flyway
- East Asian–Australian Flyway
- West Pacific Flyway

wetland site under threat selected examples

major threat from human hunters

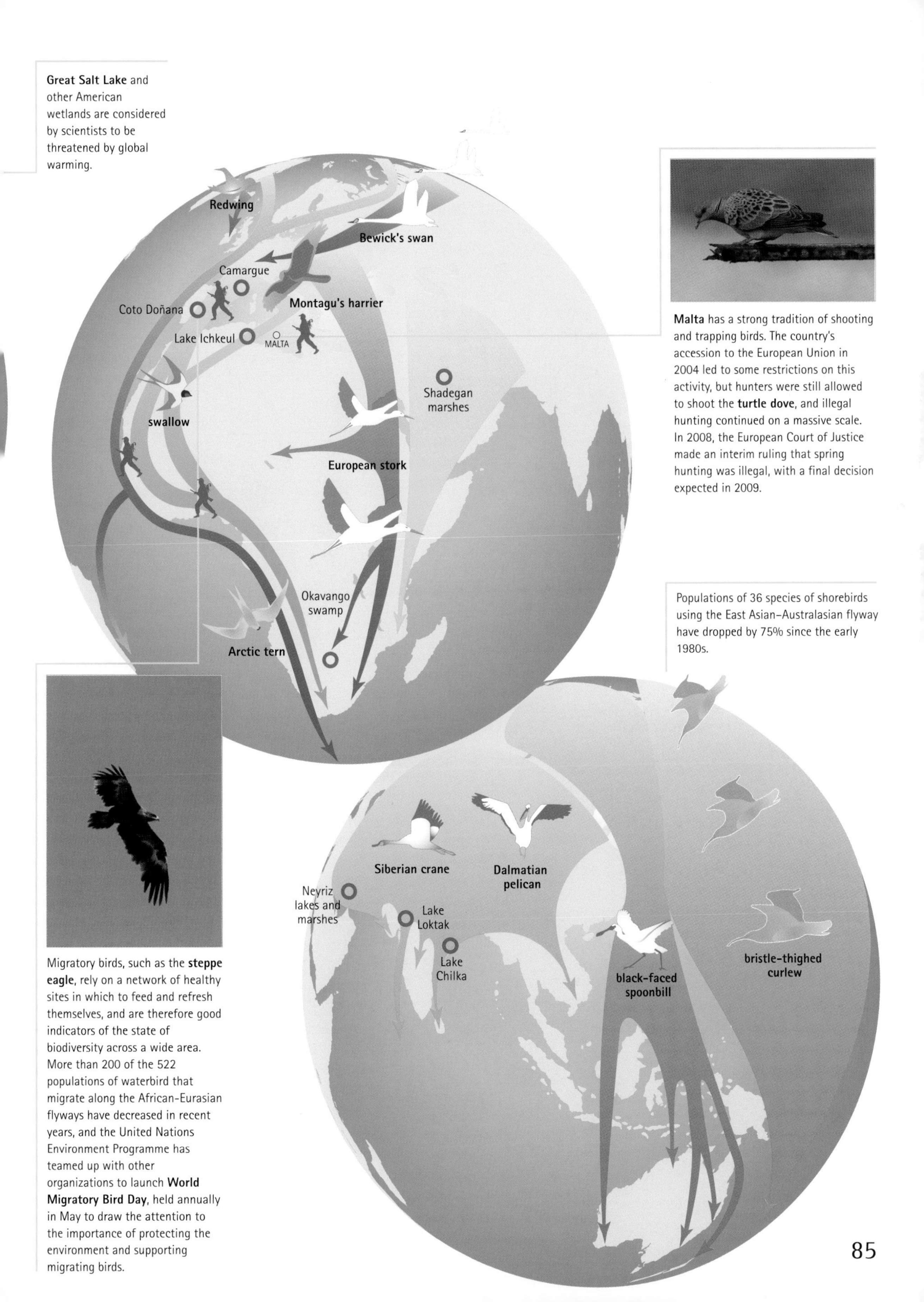

Great Salt Lake and other American wetlands are considered by scientists to be threatened by global warming.

Malta has a strong tradition of shooting and trapping birds. The country's accession to the European Union in 2004 led to some restrictions on this activity, but hunters were still allowed to shoot the **turtle dove**, and illegal hunting continued on a massive scale. In 2008, the European Court of Justice made an interim ruling that spring hunting was illegal, with a final decision expected in 2009.

Populations of 36 species of shorebirds using the East Asian–Australasian flyway have dropped by 75% since the early 1980s.

Migratory birds, such as the **steppe eagle**, rely on a network of healthy sites in which to feed and refresh themselves, and are therefore good indicators of the state of biodiversity across a wide area. More than 200 of the 522 populations of waterbird that migrate along the African-Eurasian flyways have decreased in recent years, and the United Nations Environment Programme has teamed up with other organizations to launch **World Migratory Bird Day**, held annually in May to draw the attention to the importance of protecting the environment and supporting migrating birds.

Issues of Conservation

"To achieve conservation results that are ecologically viable, it is necessary to conserve networks of key sites, migration corridors, and the ecological processes that maintain healthy ecosystems."

– WWF website, Endangered Spaces

Animal Biodiversity

There are probably about 10 million species in the kingdom of animals. About 1.3 million of them have been named and described by the scientific community, but the total number can only be estimated because much of the world has not yet been properly surveyed. New techniques are constantly being developed for examining previously uncharted regions that were thought to be barren. For example, remote-control submarines are revealing some of the remarkable species to be found in the ocean depths, and the diverse communities of bacteria, protists and fish to be found in hydro-thermal vents on the sea floor (see pages 32–33).

In lakes and oceans diversity is highest near the surface, particularly in the "photic" zone (the top 100 feet or 30 metres), where photosynthesis takes place. On land, diversity tends to be at its lowest in temperate and polar regions, possibly as a result of extinctions caused by the advance of glaciers during the recent ice-ages. In the tropics, where there is reliable moisture and warm, stable temperatures, animal diversity is much higher. Species living in these regions do not need the broad characteristics necessary for coping with varying conditions (such as drought, or winter temperatures) and adapt and diversify to suit local conditions (such as a particular type of tree, or ground cover). This leads both to a rich diversity and to a large number of endemic species – those found only in one place. The endemic species shown on the map belong to groups that have been fairly well researched, but there are an even greater, and as yet uncounted, number of insects endemic to very small patches of land.

While the rate of extinction through natural processes is estimated as less than one species a year for every million species, habitat destruction has led to a current annual extinction rate of between 1,000 and 10,000 per million species. The disappearance of some species represents lost opportunities for exploiting genetic peculiarities that could have been harnessed for medicine and agriculture. A few will be "keystone" species, whose destruction will spell doom for entire communities.

Primates, such as the **chimpanzee**, along with other large mammals, tend to reproduce more slowly than smaller mammals, birds and reptiles. This makes them more vulnerable to natural disasters and human interference.

Isolated islands may not have a large number of species, but many will be endemic. Each of the Galapagos Islands is home to its own variation of species such as the **giant tortoise**, and the finch.

Animal Biodiversity

Number of mammals per 10,000 square kilometers *2000*

- more than 100
- 51 – 100
- 21 – 50
- 1 – 20
- none or no data

ENDEMIC SPECIES

100 or more *2004*

- mammals
- birds
- reptiles

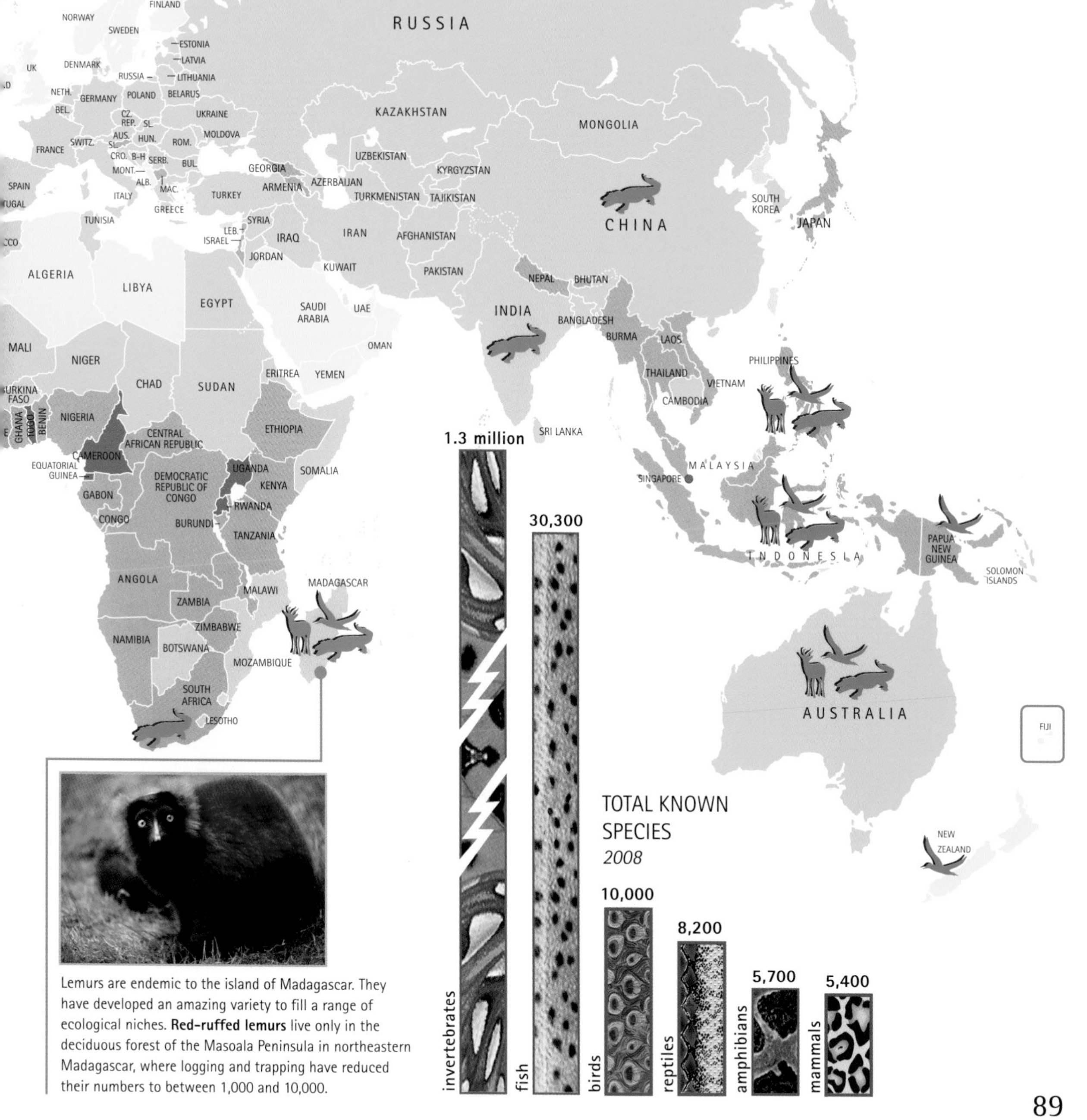

Lemurs are endemic to the island of Madagascar. They have developed an amazing variety to fill a range of ecological niches. **Red-ruffed lemurs** live only in the deciduous forest of the Masoala Peninsula in northeastern Madagascar, where logging and trapping have reduced their numbers to between 1,000 and 10,000.

Plant Biodiversity

About 300,000 species of plants have been identified, out of an estimated total of 320,000. Recently developed genetic techniques have enabled botanists to distinguish much more precisely the differences between species of plants. Flowering plants are by far the most numerous. While some are pollinated by wind, most species have arisen through successful co-evolution: using scent, colour and taste to encourage animals to pollinate them and disperse their seed.

Biodiversity is assessed in terms both of the number of species in a given area, and their abundance. A forest with one dominant species of tree and a handful of individuals of other tree species is considered to have a lower biodiveristy than a forest with the same number of tree species occurring in roughly equal numbers. As with animals, the diversity of plants tends to be highest near the equator, where the sun is strongest. In general, though, the lower and more widely fluctuating temperatures at high altitude, mean that diversity tends to be lower on mountains than at sea level.

Natural habitats provide numerous benefits, including the prevention of soil erosion, filtering of pollutants from water and the exchange of gases with the atmosphere. Such "ecosystem services" are not always taken into account, however, when development plans are drawn up. Over-extraction of natural resources and the cultivation of a small number of fast-growing varieties of crops condemn ever-larger areas of the Earth's surface to low biodiversity.

Plants have evolved to grow even in water of varying salinity. Pa-Hay-Okee ("grassy waters"), an area of 2,200 sq miles (5,700 sq km) of unique wetland habitat, was declared the **Everglades National Park** in 1947.

Epiphytes illustrate the diversity of plant form. They have no physical contact with soil, but live on the trunks and branches of other plants in order to gain access to sunlight.

The **rainforests** of Southeast Asia support a highly diverse range of plants, but are threatened with total destruction by 2050, mainly by logging but also by fire.

Plant Biodiversity

Number of Plant Species per 10,000 square kilometers *2000*

- above 4,000
- 3,001 – 4,000
- 2,001 – 3,000
- 1,001 – 2,000
- below 1,000
- none or no data
- 2,000 or more endemic vascular plants *2004*

The **Fynbos** is an area on the southern tip of Africa. About 70% of the 8,500 species of Fynbos plants are found nowhere else. Some, such as the Mountain Cypress, rely on periodic natural fires to regenerate. However, more frequent, accidental fires have disrupted this rich habitat. Much of the Lowland Fynbos has been lost to cereal fields and vineyards.

With threatened and degraded habitats on every continent, conservation organizations face a daunting task. Their strategy is to focus on key areas.

Conservation International has identified a number of "biodiversity hotspots" (see map) which have a high density of different species, a large number of endemic species, and have lost at least 70 percent of their original habitat. Although these hotspots, in total, comprise less than 2 percent of the Earth's land surface, they contain around 44 percent of all vascular plant species and 35 percent of terrestrial vertebrates.

The WWF has also identified key ecoregions around the globe on which to focus its work – 200 of them. At the beginning of 2007, however, it also launched a new program. The Protected Areas for a Living Planet (see map) aims to facilitate the implementation of the Convention on Biological Diversity (1992), which has now been signed by 190 countries and includes a commitment to create a global network of comprehensive, well-managed and representative terrestrial and marine protected areas. The six ecoregions on which work has begun involve 34 countries, and brings together partner organizations and scientific institutions as well as governments.

Wilderness areas – where 75 percent of the natural habit is undisturbed, and there is a density of no more than 13 people per square mile (5 people per sq km) – are also the focus of the attention of conservationists. Although the size of the Amazon and Congo rainforests makes them appear unassailable, the speed of development is giving cause for deep concern. If no action is taken, their ecology will become degraded through fragmentation and fire damage long before the forests have vanished completely.

Ecological conservation cannot be imposed by external organizations or even by national governments. It requires the cooperation of those living and working in the area, many of whom depend on the very activities the conservation groups are trying to stop. So, the emphasis now is on training, education and the promotion of sustainable industries.

The **giant sequoia** is one of the many species found in the **California Floristic Province**, which is also home to a number of threatened endemic species, such as the giant kangaroo rat, and desert slender salamander. It is the largest avian breeding ground in the USA.

The Atlantic forest has been reduced to around 7 percent of its original terrain, with some so fragmented as to make its survival unlikely. The **golden lion tamarin**, found nowhere else, is just one of nine critically endangered mammals in this hotspot.

Conservation Focus

2008

Conservation International biodiversity hotspots

WWF Protected Areas for a Living Planet

The **Dinaric Arc** ecoregion includes the most extensive network of subterranean rivers and lakes in Europe, the shallow reefs and marine feeding grounds of the Dalmatian Coast, and the Dinaric Alps, inhabited by lynx, brown bear, and wolf.

The **Altai-Sayan** ecoregion is one of the remaining untouched areas of the world. It includes several major mountain ranges, and a large area of forest. More than 200 rare plant species are found there, as is the critically endangered snow leopard.

Altai Sayan
European Alps
Mediterranean Basin
Carpathian
Caucasus
Mountains of Central Asia
Mountains of South-West China
Dinaric Arc
Irano-Anatolian
Himalaya
Indo-Burma
Polynesia/Micronesia
West African Forests
Eastern Afromontane
Horn of Africa
Philippines
Western Ghats and Sri Lanka
Forests of the Lower Mekong
West Africa Marine
Coastal Forests of Eastern Africa
Madagascar
Sundaland
Wallacea
East Melanesian Islands
Succulent Karoo
New Caledonia
Maputaland Pondoland-Albany
Southwest Australia
Cape Floristic Province
New Zealand

The Guinean forests of West Africa are under pressure from slash-and-burn agriculture and from small-scale mining. Large mammals such as antelope and primates are hunted for their meat and elephants are poached for ivory. The **pygmy hippo** and the white-breasted guineafowl are two of the many threatened species found in this region.

The Philippines boasts one of the most diverse ecosystems. A large proportion of its species are endemic, making the drive to conserve the remaining 8 percent of the islands' original vegetation even more urgent.

Conserving Animals

How best to save threatened animal species has been a fiercely contested subject. In an ideal world all animals would be conserved in their natural habitat, but despite the efforts of conservationists the predicted rate of habitat destruction could result in as many as 2,000 species needing support through captive breeding if they are to survive. The conservation movement is now focusing on captive breeding undertaken alongside programs to conserve and manage the habitats in which the species thrive.

In order to keep a species genetically diverse and healthy, populations of between 250 and 500 are necessary, with breeding between animals in different centres. The breeding of endangered animals is in many cases managed by dedicated international programs, which keep detailed studbooks.

There are currently around 250,000 animals in the world's zoos, but many of the species most prolifically represented are not the most threatened. Zoos are expensive to run, and although some zoos receive funding from national or local governments, most rely largely on ticket sales for their income. About 600 million people visit a zoo each year, but many zoos are faced with a dilemma – whether to provide a wide range of the more popular animals in order to attract more visitors, or to concentrate on threatened species.

Sometimes, animals bred in captivity can be released into the wild. This has been done successfully with the California Condor (see page 78), the Mallorcan midwife toad, and the Mauritius kestrel among others, but much depends on how well these released animals can be protected in the wild, and whether there is any suitable habitat left into which to release them.

The **Durrell Wildlife Conservation Trust**, based at Jersey Zoo, runs captive-breeding programs and conservation projects around the world in support of critically endangered animals. These include the **Alaotran gentle lemur**, whose habitat is reduced to a single area of wetland in Madagascar, which was declared a Ramsar Site in 2003.

The habitat of the **black lion tamarin** is Brazil's Atlantic Forest, less than a tenth of which is left. Several of the species have been successfully bred at Jersey Zoo, but have not adapted as well as hoped on release into the Brazilian forest. Conservationists are now focusing on the strategy of replanting forest to create link corridors between isolated groups of tamarins.

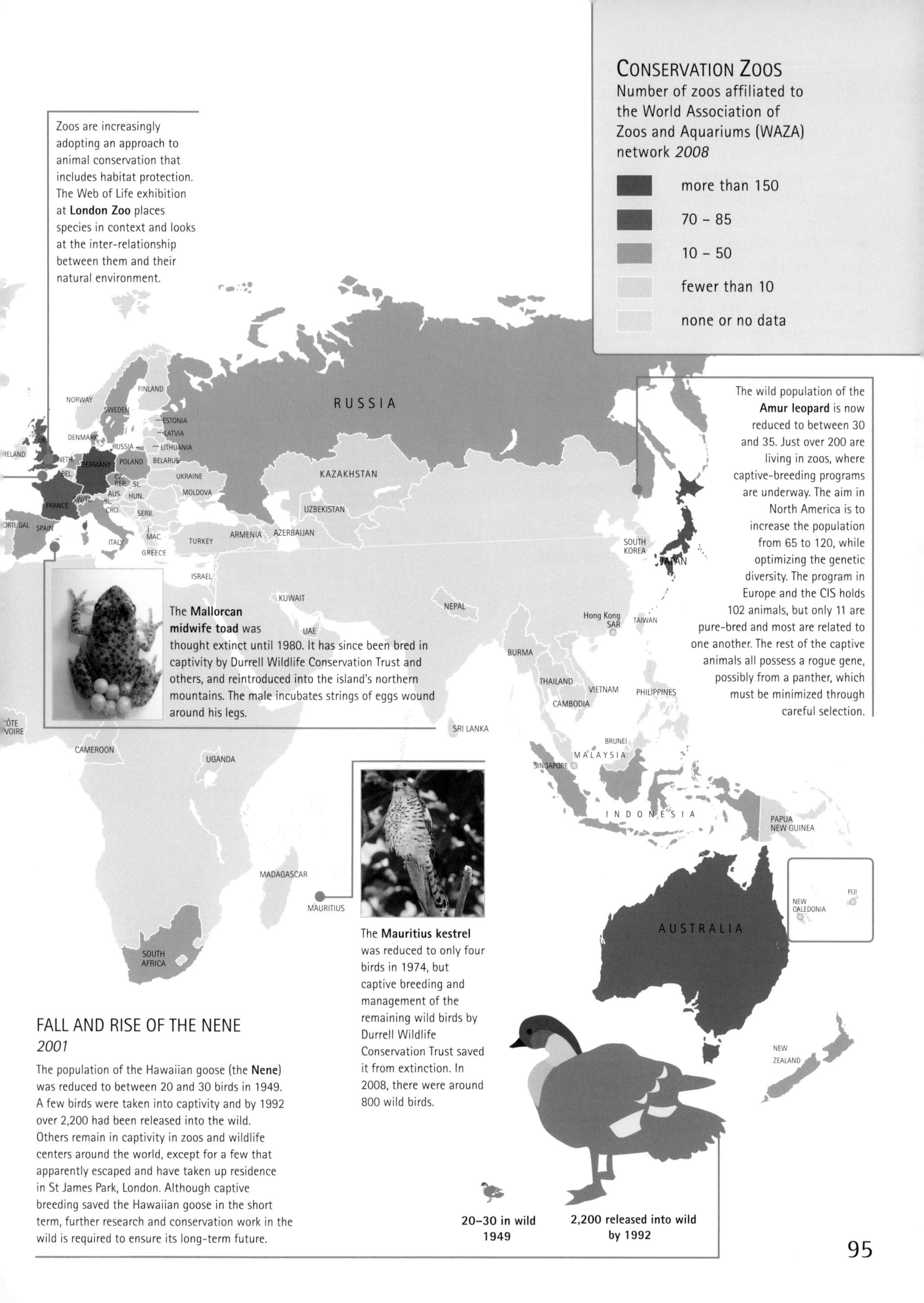

Conservation Zoos

Number of zoos affiliated to the World Association of Zoos and Aquariums (WAZA) network *2008*

- more than 150
- 70 – 85
- 10 – 50
- fewer than 10
- none or no data

Zoos are increasingly adopting an approach to animal conservation that includes habitat protection. The Web of Life exhibition at **London Zoo** places species in context and looks at the inter-relationship between them and their natural environment.

The wild population of the **Amur leopard** is now reduced to between 30 and 35. Just over 200 are living in zoos, where captive-breeding programs are underway. The aim in North America is to increase the population from 65 to 120, while optimizing the genetic diversity. The program in Europe and the CIS holds 102 animals, but only 11 are pure-bred and most are related to one another. The rest of the captive animals all possess a rogue gene, possibly from a panther, which must be minimized through careful selection.

The **Mallorcan midwife toad** was thought extinct until 1980. It has since been bred in captivity by Durrell Wildlife Conservation Trust and others, and reintroduced into the island's northern mountains. The male incubates strings of eggs wound around his legs.

The **Mauritius kestrel** was reduced to only four birds in 1974, but captive breeding and management of the remaining wild birds by Durrell Wildlife Conservation Trust saved it from extinction. In 2008, there were around 800 wild birds.

FALL AND RISE OF THE NENE

2001

The population of the Hawaiian goose (the **Nene**) was reduced to between 20 and 30 birds in 1949. A few birds were taken into captivity and by 1992 over 2,200 had been released into the wild. Others remain in captivity in zoos and wildlife centers around the world, except for a few that apparently escaped and have taken up residence in St James Park, London. Although captive breeding saved the Hawaiian goose in the short term, further research and conservation work in the wild is required to ensure its long-term future.

Conserving Plants

Botanic gardens are the zoos of the plant world. There are about 1,600 of them, containing roughly a quarter of all species of flowering plants and ferns. They are visited each year by about 150 million people worldwide and, as with zoos, public education is an important function. Some gardens incorporate herbaria, where plant specimens are dried and held to provide a permanent reference for plant diversity. Botanic gardens often also maintain seed banks. Seeds are only viable for a certain period, so drying and low-temperature storage ("cryopreservation") is used to prevent them germinating or rotting .

Governments around the world have declared nature reserves in order to protect important habitats. However, the size of a protected area is significant. Small reserves hold fewer individuals of any one plant species, which leads to a greater risk of that plant becoming extinct. Reserves need to be close to each other, or connected by similar habitat, to allow pollination and seed dispersal between them.

Preventing frequent fires is important for most plants. In the tundra the risk of fire is relatively low, plant diversity is modest and there is limited scope for agriculture. In consequence, fewer protected areas have been designated there.

Invasive plants can overwhelm ecologically balanced native flora. Physical destruction of invasive plants, pesticides and the introduction of herbivores and parasites are all used to control invasive species. Genetic pollution is a more recent hazard. Genetically modified crops could, in theory, pollinate natural relatives, with the result that genes from organisms as exotic as fish are spread to wild plants.

Eco-labelling of timber can exploit market mechanisms to promote conservation. The Forest Stewardship Council administers a certification scheme for timber extracted according to strict ecological principles. These aim either to leave forests largely intact or to give the species that are cut down an opportunity to regenerate. The success of such conservation strategies relies on harnessing the knowledge of local people about the plants around them.

Over 900 invasive species of plant, such as the **Japanese knotweed**, threaten the native flora of the USA.

Every year 12 million hectares of forest worldwide are cleared. About 8% of global forests are protected, but the loss of surrounding forests risks creating isolated fragments of habitat, which inevitably become less diverse.

PLANT CONSERVATION INSTITUTIONS

Number linked to regional association affiliated to Botanic Gardens Conservation International

2008

- over 100
- 50 – 99
- 10 – 49
- fewer than 10
- none or no data

The **Svalbard Global Seed Vault** is over 130 meters deep, and was built by the Norwegian government to house millions of seeds from around the world. Nicknamed the "Doomsday Vault", it is intended as a backup to national seed vaults in the event of natural or man-made disasters, which have been known to destroy entire collections.

China is home to an estimated 31,000 plant species. In 2007 it launched a new National Strategy for Plant Conservation to halt the country's continuing loss of plant diversity, and to protect plants such as the **Magnolia sieboldii**, common in European gardens, but vulnerable in the wild.

The **golden pagoda** was only discovered in 1987. It is limited in the wild to the mountains bordering Little Karoo in Western Cape. Although it is largely contained within a nature reserve, care is needed to ensure that it does not face competition from invasive plants, and is not destroyed by fire.

Conserving Domestic Breeds

Humans have domesticated around 40 animal species. Over 12,000 years, thousands of different breeds have developed through selective breeding of these species as well as environmental factors. In recent decades, high-yielding breeds have been favored, with the result that other breeds are endangered or extinct. Over one-fifth of domestic animal breeds are threatened by extinction. On average, two breeds are disappearing every week.

Domestic animal diversity is an important component of global biodiversity. Of the 6,300 or so breeds identified by the UN Food and Agriculture Organization (FAO), around 400 are "intensively developed": they produce high yields, but they also require a high level of maintenance, such as purpose-built shelters and specialized feed stuffs. The majority of the world's people, however, still rely on breeds that require minimum shelter and natural grazing, but do not necessarily give a high yield of milk or meat. While there is a global need to increase food production, the promotion of high-yield breeds that cannot cope with adverse local conditions is not the answer.

The preservation of a wide variety of breeds provides a genetic pool from which scientists can develop animals to meet future needs. Climate change, with its attendant impact on local environmental conditions may require the development of drought-resistant breeds. The specific genes that provide certain breeds with resistance to disease can be identified and introduced into higher-yielding breeds. As food fashions change, for health and cultural reasons, breeds can be developed to meet demand.

The FAO World Watch List assesses the viability of specific breeds within a country. Some populations have been deemed too low for survival. Others are in the "critical" or "endangered" category. This draws the attention of national organizations to threatened breeds within their borders, but it also produces anomalies. Countries in Africa, where data is not available on many breeds, register relatively few threatened breeds. On the other hand, European countries register a large number of threatened breeds because efforts are being made to conserve them.

The **angler sattelschwein** breed was reduced to only 30 breeding females in 1990, but this has since been increased to just under 100.

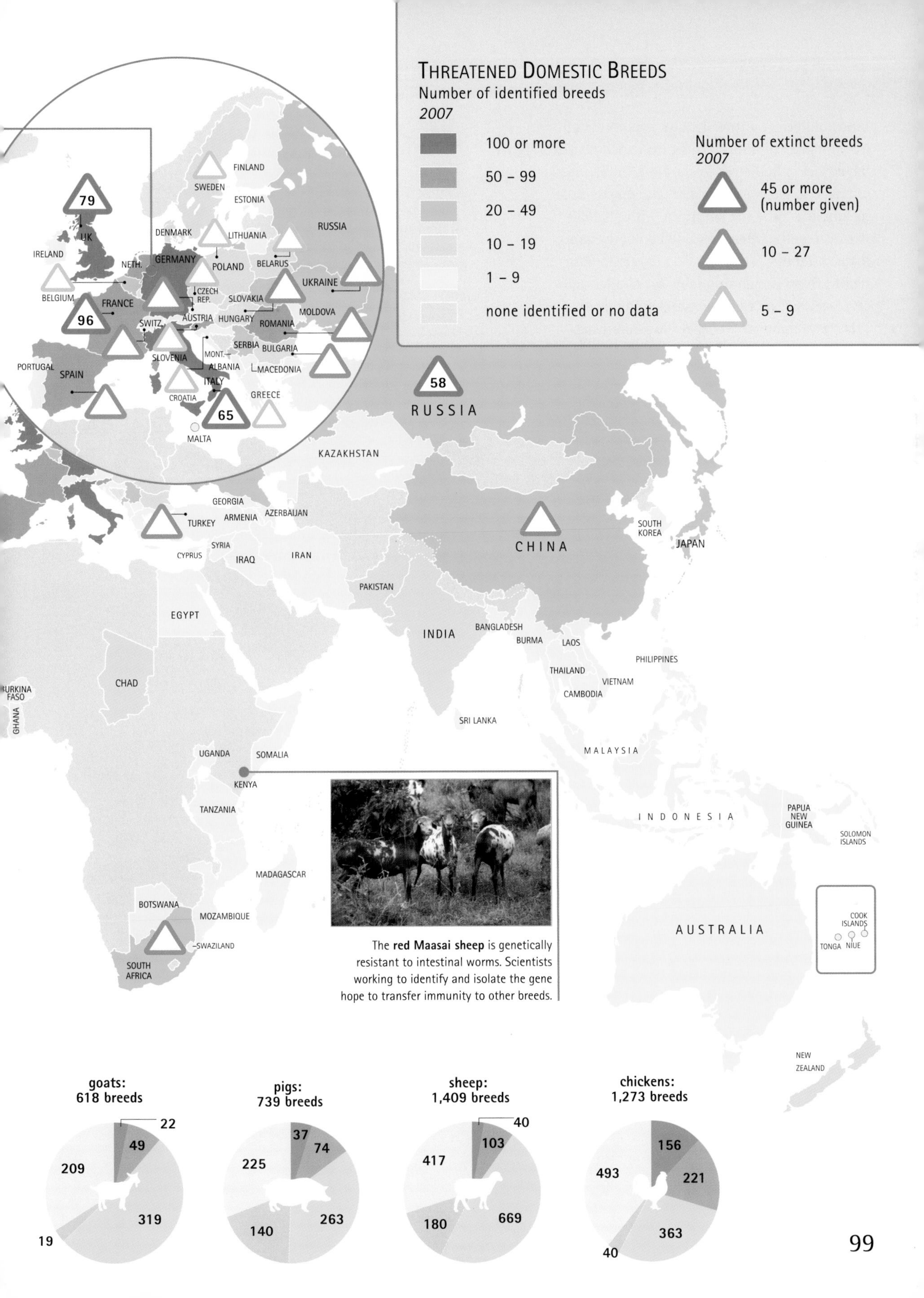

The **red Maasai sheep** is genetically resistant to intestinal worms. Scientists working to identify and isolate the gene hope to transfer immunity to other breeds.

Import Trade

Industrialized countries legally imported 38,500 live primates in 2002, largely for use in research laboratories; half went to the USA. Trade in other animals, in particular parrots and reptiles, is mostly driven by the pet trade, with 385,600 parrots legally imported in 2002, nearly half of them to Europe. There is also a large trade in reptile skins, now fashionable for clothes and accessories. Over 200,000 alligator skins are sold each year, although many of these are from farmed, rather than wild, animals. Reptile farming practices in some countries are brutal, and include skinning snakes alive, supposedly to obtain more supple skins.

The Convention on International Trade in Endangered Species of Wild Fauna and Flora (CITES) came into force in January 1975 and currently has over 170 signatory states. Under the convention, these states agree to ban commercial international trade in some species of animal and plant and to monitor trade in others. Export and import licenses have to be issued for species covered by the convention, and trade statistics reported annually to the CITES Trade Database, managed by the World Conservation Monitoring Centre in Cambridge, UK. Compliance with the reporting requirements is by no means universal, however, with many countries failing to send in even 50 percent of the required reports.

This map (and that on pages 102–103) therefore reflects legal trade, sanctioned under CITES. It does not include the billions of dollars worth of illegal trading conducted worldwide. Nor does it indicate the mortality rate among those animals licensed for live export and import. What it does show, however, is the extent of the legal trade in species considered at risk. It also identifies the major importing countries, whose demand is driving the trade.

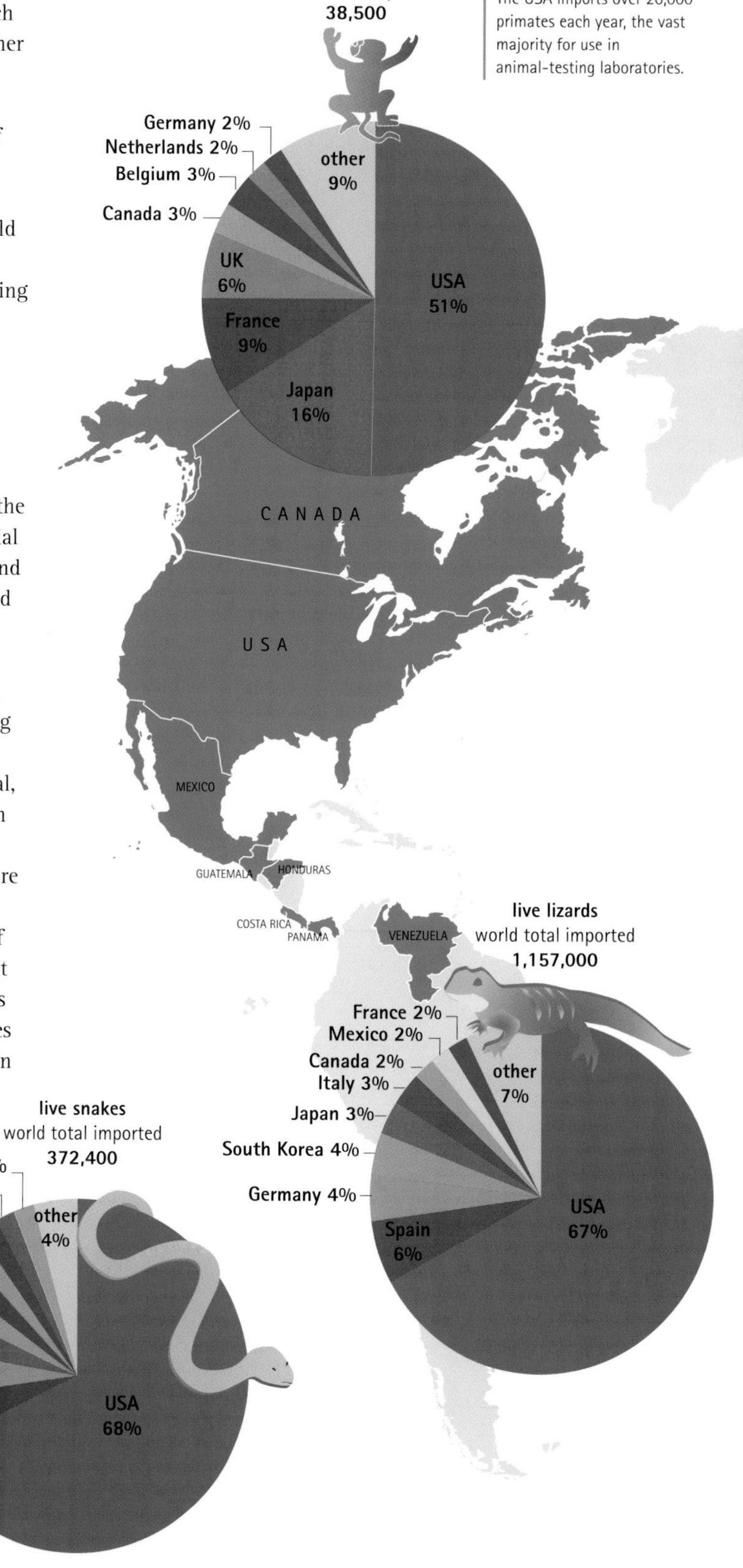

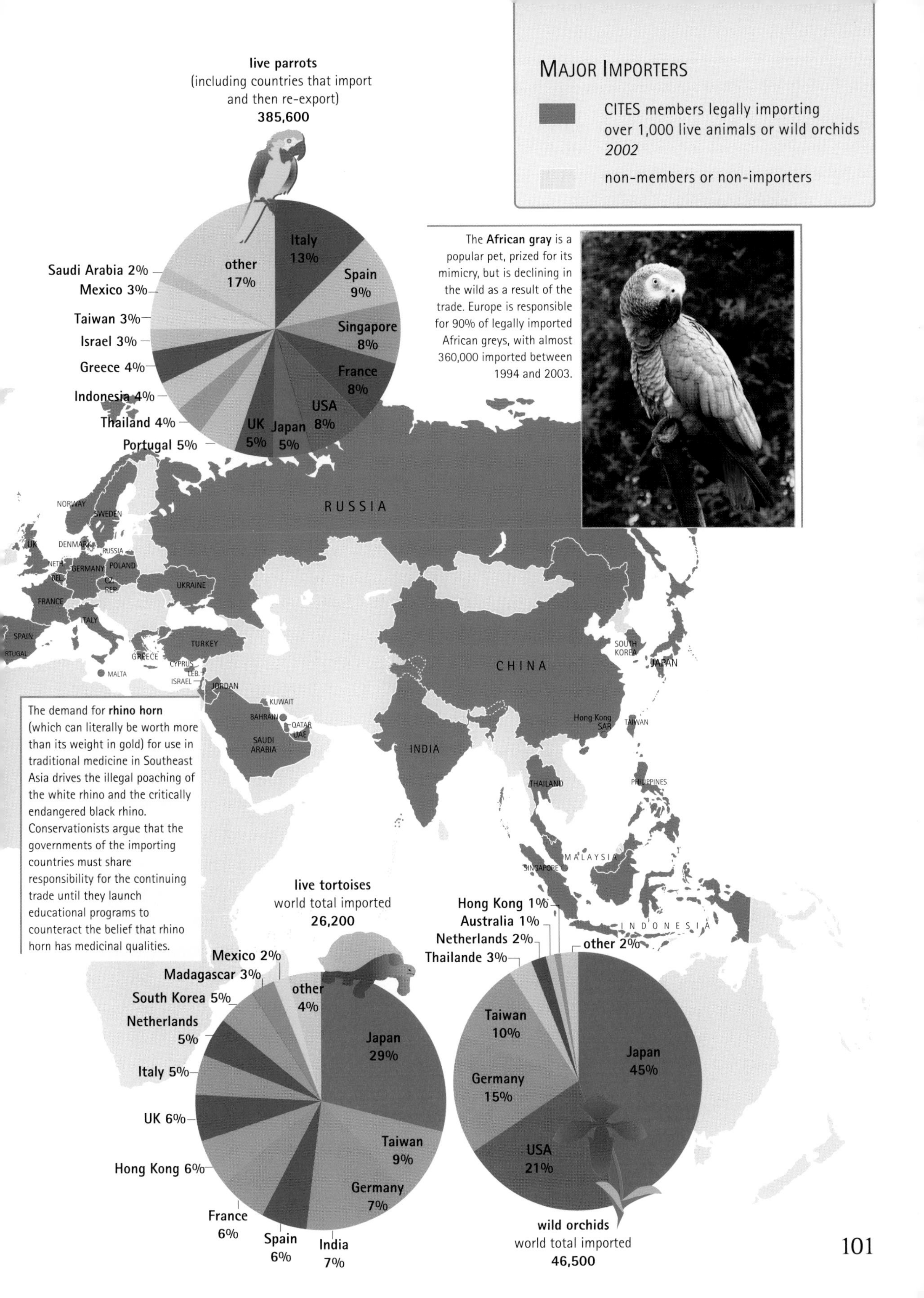

Major Importers
CITES members legally importing over 1,000 live animals or wild orchids 2002
non-members or non-importers
live parrots
(including countries that import and then re-export)
385,600
Italy 13%
Spain 9%
Singapore 8%
France 8%
USA 8%
Japan 5%
UK 5%
Portugal 5%
Thailand 4%
Indonesia 4%
Greece 4%
Israel 3%
Taiwan 3%
Mexico 3%
Saudi Arabia 2%
other 17%
The African gray is a popular pet, prized for its mimicry, but is declining in the wild as a result of the trade. Europe is responsible for 90% of legally imported African greys, with almost 360,000 imported between 1994 and 2003.
RUSSIA
NORWAY
SWEDEN
UK
DENMARK
RUSSIA
NETH.
GERMANY
POLAND
BEL.
CZ. REP.
UKRAINE
FRANCE
ITALY
SPAIN
RTUGAL
GREECE
TURKEY
MALTA
CYPRUS
LEB.
ISRAEL
JORDAN
KUWAIT
BAHRAIN
QATAR
UAE
SAUDI ARABIA
INDIA
CHINA
SOUTH KOREA
JAPAN
Hong Kong SAR
TAIWAN
THAILAND
PHILIPPINES
MALAYSIA
SINGAPORE
INDONESIA
The demand for rhino horn (which can literally be worth more than its weight in gold) for use in traditional medicine in Southeast Asia drives the illegal poaching of the white rhino and the critically endangered black rhino. Conservationists argue that the governments of the importing countries must share responsibility for the continuing trade until they launch educational programs to counteract the belief that rhino horn has medicinal qualities.
live tortoises
world total imported
26,200
Japan 29%
Taiwan 9%
Germany 7%
India 7%
Spain 6%
France 6%
Hong Kong 6%
UK 6%
Italy 5%
Netherlands 5%
South Korea 5%
Madagascar 3%
Mexico 2%
other 4%
wild orchids
world total imported
46,500
Japan 45%
USA 21%
Germany 15%
Taiwan 10%
Thailande 3%
Netherlands 2%
Australia 1%
Hong Kong 1%
other 2%

Export Trade

The financial rewards from the export of valuable wild animals and plants are considerable. Around 33,000 species are included in the Convention on International Trade in Endangered Species of Wild Fauna and Flora (CITES). For species threatened with extinction, trade is banned except in exceptional circumstances. The convention also includes species for which trade needs to be controlled in order to prevent risk of extinction. Finally, it includes species protected in at least one country, where that country has asked other CITES parties for assistance in controlling trade worldwide.

Although the Convention regularly revises the list of species it covers, it is largely the responsibility of each member state to establish the appropriate maximum number of export licenses that can be issued for a species within its territory.

Even where appropriate licensing quotas have been established, they are difficult to enforce. CITES urges countries to designate a small number of ports of exit and entry, and to train specialized enforcement officers to help identify the species being traded. Little has been done in this regard, however, although measures taken in the USA, where only nine ports are authorized to handle wildlife exports and imports, do seem to have reduced illegal trade.

The organization TRAFFIC, the trade monitoring program of WWF and IUCN-The World Conservation Union, was set up in the 1970s to assist in the implementation of CITES. In recent years it has expanded its role to look not only at trade covered by the convention but at how international sectors, including the fisheries and timber trades, impact on whole regions.

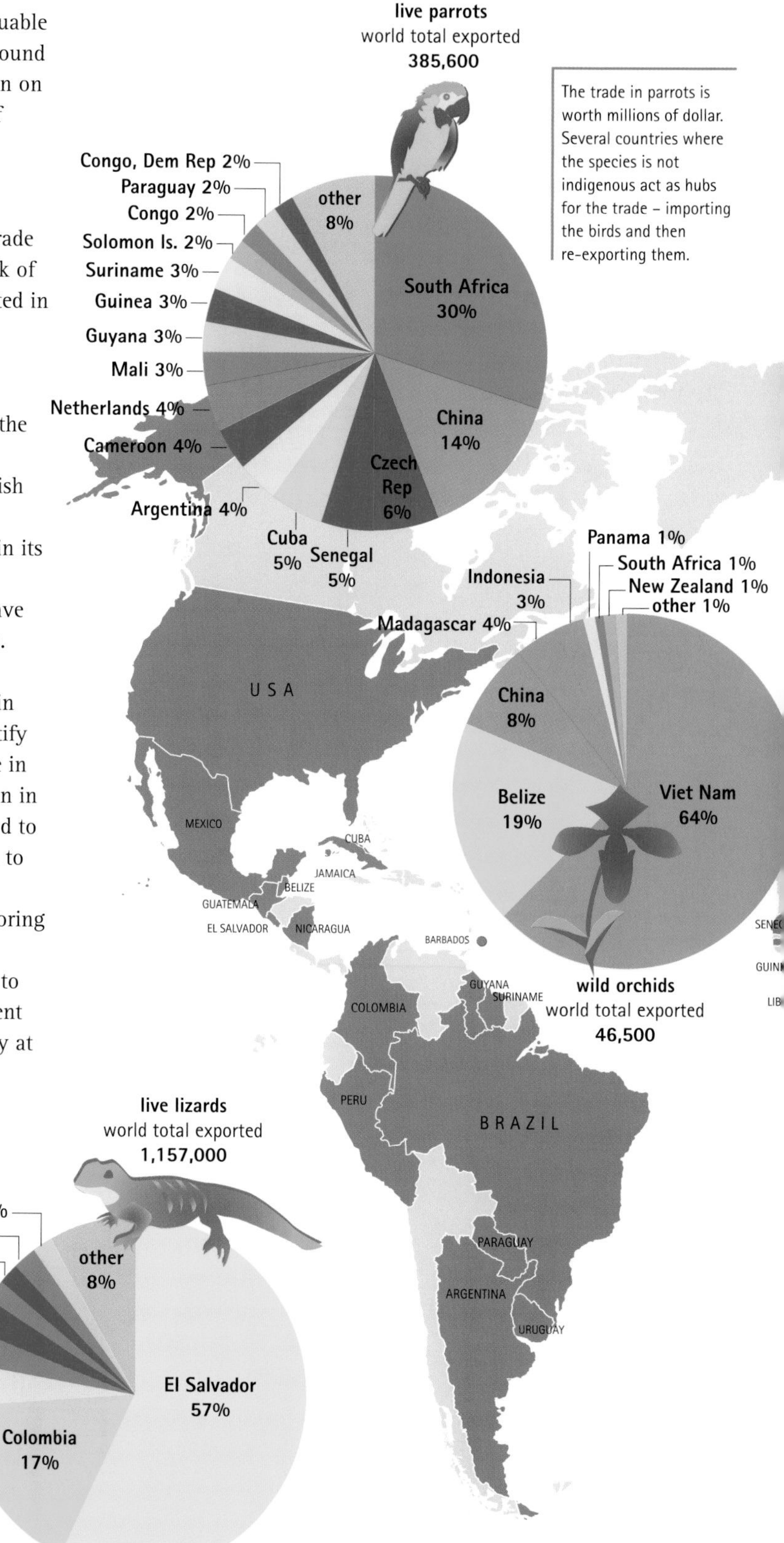

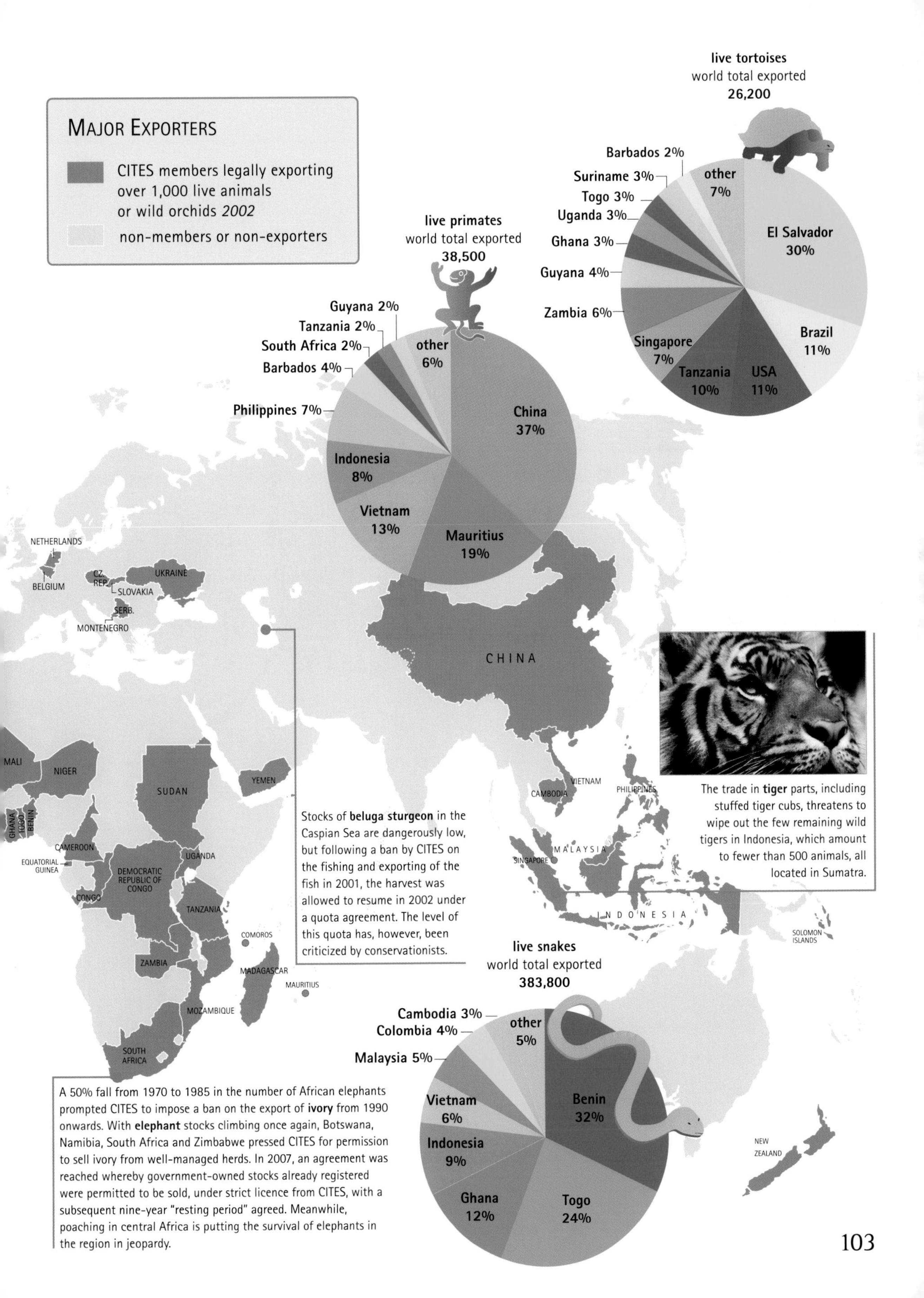

MAJOR EXPORTERS
CITES members legally exporting over 1,000 live animals or wild orchids 2002
non-members or non-exporters
live tortoises
world total exported
26,200
Barbados 2%
Suriname 3%
Togo 3%
Uganda 3%
Ghana 3%
Guyana 4%
Zambia 6%
Singapore 7%
Tanzania 10%
USA 11%
Brazil 11%
El Salvador 30%
other 7%
live primates
world total exported
38,500
Guyana 2%
Tanzania 2%
South Africa 2%
Barbados 4%
Philippines 7%
Indonesia 8%
Vietnam 13%
Mauritius 19%
China 37%
other 6%
NETHERLANDS
BELGIUM
CZ. REP.
SLOVAKIA
UKRAINE
SERB.
MONTENEGRO
CHINA
MALI
NIGER
SUDAN
YEMEN
GHANA
TOGO
BENIN
CAMEROON
EQUATORIAL GUINEA
DEMOCRATIC REPUBLIC OF CONGO
UGANDA
CONGO
TANZANIA
COMOROS
ZAMBIA
MADAGASCAR
MAURITIUS
MOZAMBIQUE
SOUTH AFRICA
VIETNAM
CAMBODIA
PHILIPPINES
MALAYSIA
SINGAPORE
INDONESIA
SOLOMON ISLANDS
NEW ZEALAND
Stocks of beluga sturgeon in the Caspian Sea are dangerously low, but following a ban by CITES on the fishing and exporting of the fish in 2001, the harvest was allowed to resume in 2002 under a quota agreement. The level of this quota has, however, been criticized by conservationists.
The trade in tiger parts, including stuffed tiger cubs, threatens to wipe out the few remaining wild tigers in Indonesia, which amount to fewer than 500 animals, all located in Sumatra.
live snakes
world total exported
383,800
Cambodia 3%
Colombia 4%
Malaysia 5%
Vietnam 6%
Indonesia 9%
Ghana 12%
Togo 24%
Benin 32%
other 5%
A 50% fall from 1970 to 1985 in the number of African elephants prompted CITES to impose a ban on the export of ivory from 1990 onwards. With elephant stocks climbing once again, Botswana, Namibia, South Africa and Zimbabwe pressed CITES for permission to sell ivory from well-managed herds. In 2007, an agreement was reached whereby government-owned stocks already registered were permitted to be sold, under strict licence from CITES, with a subsequent nine-year "resting period" agreed. Meanwhile, poaching in central Africa is putting the survival of elephants in the region in jeopardy.

World Tables

7

"For reasons of disease, genetics and simple accident, no population of wild animals can be considered secure unless it contains around 500 individuals."

– Colin Tudge,
Last Animals at the Zoo

Protected Ecosystems and Biodiversity

	1 LAND AREA	2 TROPICAL FOREST 1999 or latest available			3 TEMPERATE FOREST 1999 or latest available			5 WETLANDS
	square kilometers	sq km	sq km protected	% protected	sq km	sq km certified	% certified	sq km protected
Afghanistan	652,090	–	–	–	20,762	0	0%	–
Albania	27,400	–	–	–	10,660	0	0%	831
Algeria	2,381,740	–	–	–	26,941	0	0%	29,596
Angola	1,246,700	375,637	9,772	3%	–	–	–	–
Argentina	2,736,690	43,599	2,407	6%	190,944	0	0%	39,952
Armenia	28,200	–	–	–	3,552	0	0%	4,922
Australia	7,682,300	140,884	10,310	7%	228,767	0	0%	73,719
Austria	82,450	–	–	–	35,927	5,500	15%	1,224
Azerbaijan	82,660	–	–	–	11,333	0	0%	995
Bahamas	10,010	–	–	–	–	–	–	326
Bahrain	710	–	–	–	–	–	–	68
Bangladesh	130,170	8,625	321	4%	–	–	–	6,112
Barbados	430	4	0	0%	–	–	–	0
Belarus	207,480	–	–	–	62,801	0	0%	2,831
Belgium	30,230	–	–	–	6,872	40	1%	429
Belize	22,810	14,404	6,283	44%	–	–	–	236
Benin	110,620	15,162	2,764	18%	–	–	–	11,794
Bhutan	47,000	9,663	2,216	23%	11,293	0	0%	–
Bolivia	1,084,380	686,376	83,334	12%	–	–	–	65,181
Bosnia-Herzegovina	51,200	–	–	–	23,027	0	0%	109
Botswana	566,730	121,231	24,092	20%	–	–	–	55,374
Brazil	8,459,420	3,012,726	206,579	7%	26,129	6,660	26%	64,341
Brunei	5,270	2,479	984	40%	–	–	–	–
Bulgaria	108,640	–	–	–	37,868	0	0%	203
Burkina Faso	273,600	–	–	–	–	–	–	2,992
Burma	657,550	206,605	1,574	1%	95,735	0	0%	3
Burundi	25,680	2,190	399	18%	–	–	–	10
Cambodia	176,520	115,159	29,537	26%	–	–	–	546
Cameroon	465,400	200,088	11,991	6%	–	–	–	6,066
Canada	9,093,510	–	–	–	4,043,133	43,600	1%	130,667
Cape Verde	4,030	–	–	–	–	–	–	–
Central African Rep.	623,000	171,008	34,419	20%	–	–	–	–
Chad	1,259,200	35,161	1,266	4%	–	–	–	98,791
Chile	748,800	–	–	–	145,265	0	0%	1,592
China	9,327,490	1,089	142	13%	827,097	0	0%	29,375
Colombia	1,109,500	531,861	57,516	11%	–	–	–	4,479
Comoros	1,860	–	–	–	–	–	–	160
Congo	341,500	243,213	10,774	4%	–	–	–	4,390
Congo, Dem. Rep.	2,267,050	1,350,712	89,705	7%	–	–	–	–
Costa Rica	51,060	14,640	6,553	45%	–	–	–	5,101
Côte-d'Ivoire	318,000	27,018	6,156	23%	–	–	–	1,273
Croatia	55,920	–	–	–	13,913	1,670	12%	866
Cuba	109,820	17,614	2,701	15%	–	–	–	11,884

SPECIES DENSITY Number of species per 10,000 sq km					
6 MAMMALS	7 BIRDS	8 REPTILES	9 AMPHIBIANS	10 PLANTS	
31	59	26	2	1,008	Afghanistan
48	162	22	9	2,139	Albania
15	32	13	2	520	Algeria
56	156	–	–	1,055	Angola
50	140	37	24	1,463	Argentina
59	169	36	5	–	Armenia
29	72	83	23	1,741	Australia
41	106	7	10	1,537	Austria
49	122	26	5	2,109	Azerbaijan
–	–	–	–	–	Bahamas
–	–	–	–	–	Bahrain
45	122	49	8	2,074	Bangladesh
–	–	–	–	–	Barbados
27	81	3	4	772	Belarus
40	125	6	12	1,073	Belgium
95	271	81	24	2,200	Belize
85	138	–	–	990	Benin
59	269	11	14	3,281	Bhutan
67	–	45	26	3,885	Bolivia
42	127	16	9	–	Bosnia-Herzegovina
43	101	41	10	563	Botswana
45	162	53	63	6,058	Brazil
–	–	–	–	–	Brunei
37	108	15	8	1,615	Bulgaria
49	112	–	–	369	Burkina Faso
62	216	51	19	1,742	Burma
76	322	–	–	1,783	Burundi
47	118	32	11	–	Cambodia
114	193	51	53	2,310	Cameroon
20	44	4	4	335	Canada
–	–	–	–	–	Cape Verde
53	137	33	12	921	Central African Rep.
27	75	1	–	322	Chad
22	71	20	12	1,269	Chile
41	114	35	30	3,340	China
75	356	124	143	10,735	Colombia
–	–	–	–	–	Comoros
62	140	–	–	1,870	Congo
74	153	62	13	1,818	Congo, Dem. Rep.
120	350	125	98	7,074	Costa Rica
73	170	–	–	1,163	Côte-d'Ivoire
43	126	16	11	–	Croatia
14	62	47	25	2,949	Cuba

PROTECTED ECOSYSTEMS AND BIODIVERSITY

	1 LAND AREA	2 TROPICAL FOREST *1999 or latest available*			3 TEMPERATE FOREST *1999 or latest available*			5 WETLANDS
	square kilometers	sq km	sq km protected	% protected	sq km	sq km certified	% certified	sq km protected
Cyprus	9,240	–	–	–	1,396	0	0%	16
Czech Republic	77,260	–	–	–	24,806	100	0%	547
Denmark	42,430	–	–	–	4,587	0	0%	20,788
Djibouti	23,180	327	0	0%	–	–	–	30
Dominican Republic	48,380	11,710	1,975	17%	–	–	–	200
Ecuador	276,840	135,082	32,280	24%	–	–	–	1,708
Egypt	995,450	1,335	0	0%	44	0	0%	1,057
El Salvador	20,720	1,109	50	5%	–	–	–	1,258
Equatorial Guinea	28,050	17,486	0	0%	–	–	–	–
Eritrea	101,000	10	0	0%	–	–	–	–
Estonia	42,390	119,373	22,498	19%	15,240	0	0%	2,260
Ethiopia	1,000,000	–	–	–	–	–	–	–
Fiji	18,270	6,411	61	1%	–	–	–	6
Finland	304,590	–	–	–	253,085	219,000	87%	7,995
France	550,100	–	–	–	108,306	10	0%	8,288
Gabon	257,670	214,815	7,778	4%	–	–	–	17,638
Gambia	10,000	1,878	96	5%	–	–	–	263
Georgia	69,490	–	–	–	31,581	0	0%	345
Germany	348,770	–	–	–	104,015	32,420	31%	8,431
Ghana	227,540	16,943	1,197	7%	–	–	–	1,784
Greece	128,900	–	–	–	44,225	0	0%	1,635
Guatemala	108,430	38,622	12,310	32%	–	–	–	6,286
Guinea	245,720	30,730	331	1%	–	–	–	64,224
Guinea Bissau	28,120	11,411	0	0%	–	–	–	391
Guyana	196,850	178,444	2,358	1%	–	–	–	–
Haiti	27,560	638	13	2%	–	–	–	–
Honduras	111,890	52,733	9,656	18%	–	–	–	2,233
Hungary	89,610	–	–	–	7,765	0	0%	2,354
Iceland	100,250	–	–	–	–	–	–	590
India	2,973,190	444,499	39,496	9%	92,603	0	0%	6,771
Indonesia	1,811,570	887,437	185,342	21%	–	–	–	6,565
Iran	1,628,550	–	–	–	23,479	0	0%	14,811
Iraq	437,370	–	–	–	–	–	–	1,377
Ireland	68,890	–	–	–	4,567	0	0%	670
Israel	21,640	–	–	–	–	–	–	4
Italy	294,110	–	–	–	67,573	110	0%	598
Jamaica	10,830	3,987	818	21%	–	–	–	378
Japan	364,500	–	–	–	56,772	30	0%	1,303
Jordan	88,240	–	–	–	–	–	–	74
Kazakhstan	2,699,700	–	–	–	26,377	0	0%	3,533
Kenya	569,140	34,225	2,834	8%	–	–	–	1,018
Korea (North)	120,410	–	–	–	39,674	0	0%	–
Korea (South)	98,730	–	–	–	14,260	0	0%	46

	SPECIES DENSITY Number of species per 10,000 sq km				
	6 MAMMALS	7 BIRDS	8 REPTILES	9 AMPHIBIANS	10 PLANTS
Cyprus	–	–	–	–	–
Czech Republic	41	101	5	10	–
Denmark	27	121	3	9	895
Djibouti	–	–	–	–	–
Dominican Republic	12	81	69	21	3,354
Ecuador	100	460	126	141	6,421
Egypt	21	33	18	1	454
El Salvador	106	196	57	18	2,277
Equatorial Guinea	131	194	–	–	2,312
Eritrea	50	141	38	8	–
Estonia	40	130	3	7	1,018
Ethiopia	54	133	40	13	1,398
Fiji	3	61	20	2	1,334
Finland	19	78	2	2	345
France	25	72	9	9	1,233
Gabon	64	157	–	–	2,248
Gambia	112	269	45	29	935
Georgia	56	–	27	7	2,292
Germany	23	73	4	6	824
Ghana	78	186	–	–	1,308
Greece	41	107	24	6	2,131
Guatemala	114	208	107	49	3,948
Guinea	66	142	–	–	1,043
Guinea Bissau	71	159	–	–	655
Guyana	70	246	–	–	2,329
Haiti	2	54	77	40	3,743
Honduras	78	190	73	36	2,559
Hungary	40	98	7	8	1,155
Iceland	5	41	0	0	175
India	47	137	58	31	2,363
Indonesia	81	271	91	50	5,196
Iran	26	60	31	2	1,489
Iraq	23	49	23	2	–
Ireland	13	75	1	2	499
Israel	91	141	76	5	2,194
Italy	29	76	13	13	1,820
Jamaica	23	110	35	23	3,207
Japan	57	75	26	18	1,679
Jordan	34	68	35	–	1,069
Kazakhstan	28	62	8	2	–
Kenya	94	222	50	23	1,703
Korea (North)	–	51	8	6	1,274
Korea (South)	23	53	12	7	1,359

Protected Ecosystems and Biodiversity

	1 LAND AREA	2 TROPICAL FOREST 1999 or latest available			3 TEMPERATE FOREST 1999 or latest available			5 WETLANDS sq km protected
	square kilometers	sq km	sq km protected	% protected	sq km	sq km certified	% certified	
Kuwait	17,820	–	–	–	–	–	–	–
Kyrgyzstan	191,800	–	–	–	7,846	0	0%	6,397
Laos	230,800	36,392	8,388	23%	8,492	0	0%	–
Latvia	62,290	–	–	–	16,239	0	0%	1,484
Lebanon	10,230	–	–	–	355	0	0%	11
Lesotho	30,350	889	78	9%	–	–	–	4
Liberia	96,320	31,485	913	3%	–	–	–	959
Libya	1,759,540	–	–	–	525	0	0%	1
Lithuania	62,680	–	–	–	15,095	0	0%	505
Luxembourg	2,590	–	–	–	806	0	0%	3
Macedonia	25,430	–	–	–	10,907	0	0%	216
Madagascar	581,540	69,401	3,831	6%	–	–	–	7,876
Malawi	94,080	38,301	3,262	9%	–	–	–	2,248
Malaysia	328,550	130,065	15,181	12%	–	–	–	554
Maldives	300	–	–	–	–	–	–	–
Mali	1,220,190	61,320	1,417	2%	–	–	–	41,195
Mauritania	1,030,700	–	–	–	–	–	–	12,311
Mauritius	2,030	–	–	–	–	–	–	4
Mexico	1,943,950	457,649	19,518	4%	212,926	1,690	1%	59,526
Moldova	32,870	–	–	–	1,434	0	0%	947
Mongolia	1,566,500	–	–	–	26,365	0	0%	14,395
Montenegro	13,810	–	–	–	36,644	0	0%	200
Morocco	446,300	–	–	–	18,621	0	0%	2,720
Mozambique	786,380	208,630	15,544	7%	–	–	–	6,880
Namibia	823,290	34,362	3,626	11%	–	–	–	6,296
Nepal	143,000	11,622	2,184	19%	26,605	0	0%	344
Netherlands	33,880	–	–	–	2,351	690	29%	8,189
New Zealand	267,710	–	–	–	42,116	3,630	9%	391
Nicaragua	121,400	53,225	13,134	25%	–	–	–	4,057
Niger	1,266,700	269	42	16%	–	–	–	43,176
Nigeria	910,770	116,338	8,583	7%	–	–	–	10,767
Norway	304,280	–	–	–	81,386	56,000	69%	1,164
Oman	309,500	–	–	–	–	–	–	–
Pakistan	770,880	8,065	45	1%	20,830	0	0%	13,436
Panama	74,430	37,444	11,577	31%	–	–	–	1,599
Papua New Guinea	452,860	357,905	38,257	11%	–	–	–	5,949
Paraguay	397,300	92,904	2,440	3%	28,476	0	0%	7,860
Peru	1,280,000	756,362	38,917	5%	–	–	–	67,806
Philippines	298,170	24,023	1,248	5%	–	–	–	684
Poland	306,330	–	–	–	89,388	27,430	31%	1,451
Portugal	91,500	–	–	–	26,613	0	0%	738
Qatar	11,000	–	–	–	–	–	–	–
Romania	229,980	–	–	–	81,372	0	0%	6,836

	SPECIES DENSITY Number of species per 10,000 sq km				
6 MAMMALS	7 BIRDS	8 REPTILES	9 AMPHIBIANS	10 PLANTS	
17	17	24	2	193	Kuwait
31	–	12	1	1,412	Kyrgyzstan
61	171	23	13	–	Laos
45	117	4	7	651	Latvia
56	152	41	8	2,961	Lebanon
23	40	–	–	1,103	Lesotho
87	168	28	17	993	Liberia
14	17	10	1	331	Libya
37	109	4	7	967	Lithuania
–	–	–	–	–	Luxembourg
57	154	23	10	2,563	Macedonia
37	53	95	47	2,479	Madagascar
86	230	55	31	1,665	Malawi
95	160	110	60	4,890	Malaysia
–	–	–	–	–	Maldives
28	81	3	–	355	Mali
13	59	–	–	239	Mauritania
–	–	–	–	–	Mauritius
86	135	123	54	4,569	Mexico
46	119	6	9	1,173	Moldova
25	80	4	1	533	Mongolia
45	104	33	10	1,896	Montenegro
30	60	26	3	1,049	Morocco
42	117	39	15	1,340	Mozambique
58	109	58	12	942	Namibia
75	252	41	18	2,871	Nepal
35	120	4	10	767	Netherlands
1	51	18	1	802	New Zealand
86	207	69	25	3,256	Nicaragua
27	60	–	–	238	Niger
62	135	30	24	1,059	Nigeria
17	77	2	2	544	Norway
20	39	23	–	439	Oman
36	88	41	4	1,168	Pakistan
112	376	116	84	5,088	Panama
63	184	79	63	3,257	Papua New Guinea
90	164	35	25	2,311	Paraguay
93	310	73	76	3,674	Peru
51	64	62	30	2,907	Philippines
27	72	3	6	778	Poland
30	100	14	8	2,428	Portugal
–	–	–	–	–	Qatar
29	87	9	7	1,194	Romania

Protected Ecosystems and Biodiversity

	1 LAND AREA	2 TROPICAL FOREST *1999 or latest available*			3 TEMPERATE FOREST *1999 or latest available*			5 WETLANDS
	square kilometers	sq km	sq km protected	% protected	sq km	sq km certified	% certified	sq km protected
Russia	16,381,390	–	–	–	8,155,509	330	0%	103,238
Rwanda	24,670	2,907	2,237	77%	–	–	–	–
Samoa	2,830	–	–	–	–	–	–	–
Saudi Arabia	2,149,690	–	–	–	–	–	–	–
Senegal	192,530	20,763	1,449	7%	–	–	–	997
Serbia	77,470	–	–	–	36,644	0	0%	537
Seychelles	460	–	–	–	–	–	–	1
Sierra Leone	71,620	2,598	528	20%	–	–	–	2,950
Singapore	690	–	–	–	–	–	–	–
Slovakia	48,100	–	–	–	23,079	0	0%	407
Slovenia	20,140	–	–	–	6,957	0	0%	82
Solomon Islands	27,990	–	–	–	–	–	–	–
Somalia	627,340	118,003	1,351	1%	–	–	–	–
South Africa	1,214,470	103,326	5,334	5%	523	8,280	–	5,440
Spain	499,190	–	–	–	140,236	0	0%	2,818
Sri Lanka	64,630	15,806	4,366	28%	–	–	–	85
Sudan	2,376,000	122,882	15,065	12%	–	–	–	77,846
Suriname	156,000	132,194	5,226	4%	–	–	–	120
Swaziland	17,200	2,861	92	3%	–	–	–	–
Sweden	410,330	–	–	–	293,636	111,670	38%	5,145
Switzerland	40,000	–	–	–	13,086	490	4%	87
Syria	183,780	–	–	–	471	0	0%	100
Tajikistan	139,960	–	–	–	–	–	–	946
Tanzania	885,800	143,562	22,728	16%	–	–	–	48,684
Thailand	510,890	162,370	50,651	31%	3,608	0	0%	3,706
Togo	54,390	2,240	59	3%	–	–	–	12,104
Trinidad & Tobago	5,130	1,241	87	7%	–	–	–	159
Tunisia	155,360	–	–	–	3,005	0	0%	7,265
Turkey	769,630	–	–	–	83,898	0	0%	1,795
Turkmenistan	469,930	–	–	–	2,164	0	0%	–
Uganda	197,100	37,724	6,405	17%	–	–	–	3,548
Ukraine	579,380	–	–	–	70,458	2,030	3%	7,447
United Arab Emirates	83,600	–	–	–	–	–	–	–
United Kingdom	241,930	–	–	–	23,033	9,580	42%	9,180
United States of America	9,161,920	4,428	297	7%	2,793,863	261,290	9%	13,123
Uruguay	175,020	20	0	0%	969	0	0%	4,249
Uzbekistan	425,400	–	–	–	2,309	0	0%	313
Venezuela	882,050	556,152	328,013	59%	–	–	–	2,636
Vietnam	310,070	42,180	4,355	10%	7,233	0	0%	258
Western Sahara	266,000	–	–	–	–	–	–	–
Yemen	527,970	–	–	–	–	–	–	–
Zambia	743,390	219,885	70,176	32%	–	–	–	40,305
Zimbabwe	386,850	153,967	18,732	12%	–	–	–	–

SPECIES DENSITY Number of species per 10,000 sq km					
6 MAMMALS	7 BIRDS	8 REPTILES	9 AMPHIBIANS	10 PLANTS	
23	58	5	4	–	Russia
110	373	–	–	1,664	Rwanda
–	–	–	–	–	Samoa
13	26	14	–	345	Saudi Arabia
72	144	37	1	780	Senegal
45	104	33	10	1,896	Serbia
–	–	–	–	–	Seychelles
77	243	–	–	1,091	Sierra Leone
213	295	350	60	5,713	Singapore
50	124	12	12	1,849	Slovakia
59	164	20	16	2,535	Slovenia
37	115	43	12	2,235	Solomon Islands
43	107	49	7	768	Somalia
52	122	65	22	4,797	South Africa
22	76	15	8	1,383	Spain
47	134	77	21	1,781	Sri Lanka
43	110	–	–	507	Sudan
72	240	60	38	1,997	Suriname
–	–	–	–	–	Swaziland
17	71	2	4	498	Sweden
47	121	9	11	1,898	Switzerland
24	78	–	–	1,145	Syria
35	–	18	1	–	Tajikistan
70	184	64	30	2,231	Tanzania
72	168	81	31	3,170	Thailand
110	220	–	–	1,739	Togo
125	324	87	32	2,816	Trinidad & Tobago
31	69	25	3	873	Tunisia
28	72	25	4	2,059	Turkey
29	–	23	1	–	Turkmenistan
118	290	52	17	1,891	Uganda
28	68	5	4	1,318	Ukraine
12	33	18	–	–	United Arab Emirates
17	80	3	2	565	United Kingdom
45	68	30	28	2,036	United States of America
31	92	–	–	882	Uruguay
28	–	18	1	1,369	Uzbekistan
73	302	64	55	4,752	Venezuela
67	168	59	25	3,306	Vietnam
–	–	–	–	–	Western Sahara
18	39	21	–	446	Yemen
56	145	35	16	1,141	Zambia
81	159	46	36	1,325	Zimbabwe

Threatened Species

	Number of Threatened Species 2008					
	1 PRIMATES	2 CATS	3 UNGULATES	4 RODENTS	5 BATS	6 DOLPHINS AND WHALES
Afghanistan	–	5	5	1	4	–
Albania	–	–	–	–	1	–
Algeria	1	5	7	–	4	–
Angola	2	3	2	–	–	2
Argentina	–	2	5	7	3	5
Armenia	–	–	4	2	5	–
Australia	–	–	–	16	8	5
Austria	–	–	–	2	3	–
Azerbaijan	–	–	5	–	5	–
Bahamas	–	–	–	1	1	1
Bahrain	–	–	1	–	–	–
Bangladesh	5	4	5	2	–	4
Barbados	–	–	–	–	–	–
Belarus	–	–	1	–	3	–
Belgium	–	–	–	1	4	3
Belize	1	–	1	–	1	–
Benin	3	3	1	–	–	1
Bhutan	2	5	6	2	–	–
Bolivia	2	1	4	6	1	1
Bosnia-Herzegovina	–	–	–	2	6	–
Botswana	–	3	1	–	1	–
Brazil	21	–	2	14	14	4
Brunei	3	5	1	–	–	–
Bulgaria	–	–	–	4	6	1
Burkina Faso	1	2	3	–	–	–
Burma	5	6	9	6	2	3
Burundi	1	3	1	2	–	–
Cambodia	6	5	6	3	–	–
Cameroon	9	3	2	9	6	–
Canada	–	1	–	2	1	9
Cape Verde	–	–	–	–	–	3
Central African Rep.	2	3	2	–	2	–
Chad	–	2	7	–	–	–
Chile	–	2	2	4	3	4
China	12	11	20	16	2	3
Colombia	7	–	3	4	11	2
Comoros	1	–	–	–	1	–
Congo	6	2	1	–	2	1
Congo, Dem. Rep.	4	3	1	4	6	1
Costa Rica	1	–	1	2	2	1
Côte-d'Ivoire	4	2	4	1	7	–
Croatia	–	–	–	1	6	–
Cuba	–	–	–	6	2	1

Number of Threatened Species 2008						
7 REPTILES	8 AMPHIBIANS	9 INVERTEBRATES	10 FISH	11 BIRDS	12 PLANTS	
1	1	1	0	14	2	Afghanistan
4	2	4	27	6	0	Albania
7	3	14	22	10	3	Algeria
4	0	5	22	18	26	Angola
5	29	10	30	49	42	Argentina
5	0	6	1	12	1	Armenia
38	47	282	87	50	55	Australia
1	0	43	7	5	4	Austria
5	0	4	5	13	0	Azerbaijan
6	0	1	20	5	5	Bahamas
4	0	0	6	4	0	Bahrain
21	1	0	12	26	12	Bangladesh
4	0	0	14	1	2	Barbados
0	0	8	0	3	0	Belarus
0	0	12	8	1	1	Belgium
5	6	1	24	3	30	Belize
4	0	0	15	4	14	Benin
1	1	1	0	16	7	Bhutan
2	21	1	0	31	71	Bolivia
2	1	10	28	6	1	Bosnia-Herzegovina
0	0	0	2	8	0	Botswana
22	110	35	66	122	382	Brazil
4	3	0	7	21	99	Brunei
2	0	7	13	12	0	Bulgaria
1	0	0	0	4	2	Burkina Faso
22	0	2	16	39	38	Burma
0	6	5	18	8	2	Burundi
11	3	0	17	24	31	Cambodia
2	53	2	43	15	355	Cameroon
3	1	12	26	18	1	Canada
0	0	0	18	4	2	Cape Verde
1	0	0	0	5	15	Central African Rep.
1	0	1	0	6	2	Chad
0	20	2	18	33	39	Chile
31	85	6	60	86	446	China
15	209	2	31	87	222	Colombia
2	0	4	5	9	5	Comoros
1	0	4	15	3	35	Congo
3	13	26	25	31	65	Congo, Dem. Rep.
8	62	12	20	17	111	Costa Rica
3	13	1	19	12	105	Côte-d'Ivoire
2	2	14	42	11	1	Croatia
7	47	5	28	17	163	Cuba

THREATENED SPECIES

	Number of Threatened Species 2008					
	1 PRIMATES	2 CATS	3 UNGULATES	4 RODENTS	5 BATS	6 DOLPHINS AND WHALES
Cyprus	–	–	2	–	1	–
Czech Republic	–	–	–	2	4	–
Denmark	–	–	–	–	2	1
Djibouti	–	1	5	–	–	–
Dominican Republic	–	–	–	2	–	1
Ecuador	2	–	2	7	8	4
Egypt	–	3	9	–	–	–
El Salvador	–	–	1	–	2	–
Equatorial Guinea	8	1	1	2	1	–
Eritrea	–	2	5	1	1	1
Estonia	–	–	–	–	1	1
Ethiopia	–	2	11	8	7	–
Fiji	–	–	–	–	4	1
Finland	–	–	–	1	–	1
France	–	–	1	1	7	3
Gabon	5	2	1	–	1	–
Gambia	2	2	2	–	–	–
Georgia	–	–	5	–	5	1
Germany	–	–	–	2	4	2
Ghana	4	2	2	2	5	–
Greece	–	–	1	4	4	–
Guatemala	1	–	1	1	2	–
Guinea	2	2	3	–	8	–
Guinea Bissau	2	2	2	–	–	–
Guyana	–	–	1	1	4	1
Haiti	–	–	–	2	–	–
Honduras	–	–	1	2	3	–
Hungary	–	–	–	2	6	–
Iceland	–	–	–	–	–	6
India	7	10	19	16	10	6
Indonesia	13	9	8	47	34	5
Iran	–	6	4	5	7	–
Iraq	–	2	3	–	4	1
Ireland	–	–	–	–	–	3
Israel	–	1	5	3	4	–
Italy	–	–	2	1	6	2
Jamaica	–	–	–	1	3	–
Japan	–	1	–	5	12	7
Jordan	–	2	5	–	3	–
Kazakhstan	–	3	4	5	2	–
Kenya	1	3	5	4	5	5
Korea (North)	–	3	2	–	1	6
Korea (South)	–	1	2	–	1	6

Number of Threatened Species 2008						
7 REPTILES	8 AMPHIBIANS	9 INVERTEBRATES	10 FISH	11 BIRDS	12 PLANTS	
4	0	0	11	4	7	Cyprus
0	0	18	9	5	4	Czech Republic
0	0	11	11	3	3	Denmark
0	0	0	14	7	2	Djibouti
10	31	6	15	14	30	Dominican Republic
10	163	51	15	68	1,838	Ecuador
11	0	1	23	10	2	Egypt
6	9	0	7	4	26	El Salvador
3	4	0	13	5	63	Equatorial Guinea
6	0	0	13	7	3	Eritrea
0	0	4	2	3	0	Estonia
1	9	15	2	21	22	Ethiopia
6	1	3	10	10	66	Fiji
0	0	10	2	3	1	Finland
5	2	63	27	5	7	France
2	3	0	21	5	108	Gabon
1	0	0	16	5	4	Gambia
7	1	9	8	8	0	Georgia
0	0	30	16	4	12	Germany
3	10	0	17	8	117	Ghana
5	5	14	50	10	11	Greece
14	76	7	18	11	84	Guatemala
1	5	3	19	12	22	Guinea
1	0	0	18	1	4	Guinea Bissau
6	6	1	23	3	22	Guyana
9	46	3	16	13	29	Haiti
11	55	1	19	7	110	Honduras
1	0	26	10	9	1	Hungary
0	0	0	11	0	0	Iceland
25	63	22	39	75	247	India
27	33	31	111	116	386	Indonesia
8	4	5	16	18	1	Iran
2	1	2	6	18	0	Iraq
0	0	3	8	1	1	Ireland
10	0	12	30	12	0	Israel
5	6	58	31	7	19	Italy
8	17	5	16	10	209	Jamaica
11	20	43	40	39	12	Japan
5	0	4	14	8	0	Jordan
2	1	4	9	20	16	Kazakhstan
5	6	32	70	27	103	Kenya
0	1	2	11	19	3	Korea (North)
0	1	2	14	27	0	Korea (South)

Threatened Species

	Number of Threatened Species 2008					
	1 PRIMATES	2 CATS	3 UNGULATES	4 RODENTS	5 BATS	6 DOLPHINS AND WHALES
Kuwait	–	2	2	–	–	–
Kyrgyzstan	–	3	2	2	1	–
Laos	9	5	7	3	1	–
Latvia	–	–	–	1	2	1
Lebanon	–	2	3	–	3	–
Lesotho	–	1	–	1	1	–
Liberia	3	2	4	–	6	1
Libya	–	3	5	–	1	–
Lithuania	–	2	1	1	2	1
Luxembourg	–	–	–	1	2	–
Macedonia	–	–	–	3	6	2
Madagascar	20	–	–	5	6	–
Malawi	–	–	1	1	–	–
Malaysia	5	7	4	8	7	3
Maldives	–	–	–	–	–	–
Mali	1	2	6	–	1	–
Mauritania	–	2	5	–	–	1
Mauritius	–	–	–	–	3	–
Mexico	1	2	1	33	15	7
Moldova	–	–	–	1	3	–
Mongolia	–	1	6	3	–	–
Montenegro	–	–	–	1	4	–
Morocco	1	5	5	1	5	2
Mozambique	–	2	1	1	2	3
Namibia	–	3	2	–	–	2
Nepal	1	6	10	4	2	1
Netherlands	–	–	–	1	4	4
New Zealand	–	–	–	–	2	5
Nicaragua	–	–	1	–	1	1
Niger	–	3	7	–	3	–
Nigeria	10	2	5	4	–	–
Norway	–	–	–	–	1	7
Oman	–	1	6	1	3	–
Pakistan	–	6	6	3	1	4
Panama	1	–	1	3	1	3
Papua New Guinea	–	–	–	15	21	–
Paraguay	–	–	3	1	1	–
Peru	4	1	3	6	15	3
Philippines	–	–	5	22	11	–
Poland	–	–	1	3	4	4
Portugal	–	1	–	1	7	4
Qatar	–	–	–	–	–	–
Romania	–	–	–	6	6	1

	Number of Threatened Species 2008					
	7 REPTILES	8 AMPHIBIANS	9 INVERTEBRATES	10 FISH	11 BIRDS	12 PLANTS
Kuwait	1	0	0	10	8	0
Kyrgyzstan	2	0	3	0	8	14
Laos	11	4	0	6	22	21
Latvia	0	0	10	4	4	0
Lebanon	6	0	3	14	6	0
Lesotho	0	0	2	1	5	1
Liberia	3	4	2	19	11	46
Libya	5	0	0	13	4	1
Lithuania	0	0	6	4	4	0
Luxembourg	0	0	4	0	0	0
Macedonia	2	0	5	8	10	0
Madagascar	20	55	32	73	35	280
Malawi	0	5	16	101	12	14
Malaysia	21	46	21	47	40	686
Maldives	2	0	0	11	0	0
Mali	1	0	0	1	6	6
Mauritania	2	0	1	22	8	0
Mauritius	7	0	32	11	11	88
Mexico	95	198	40	115	59	261
Moldova	1	0	4	9	9	0
Mongolia	0	0	3	1	20	0
Montenegro	2	1	11	20	9	0
Morocco	9	2	9	29	10	2
Mozambique	5	3	5	45	21	46
Namibia	3	1	0	20	21	24
Nepal	6	3	0	0	31	7
Netherlands	0	0	6	9	1	0
New Zealand	12	4	14	16	70	21
Nicaragua	8	10	5	22	9	39
Niger	0	0	1	2	5	2
Nigeria	3	13	1	21	12	171
Norway	0	0	9	11	2	2
Oman	4	0	4	21	9	6
Pakistan	9	0	0	20	26	2
Panama	7	55	2	21	19	194
Papua New Guinea	9	10	12	38	31	142
Paraguay	2	0	0	0	28	10
Peru	6	80	2	10	94	274
Philippines	9	48	20	58	67	213
Poland	0	0	16	4	5	4
Portugal	2	0	83	39	8	16
Qatar	1	0	0	7	4	0
Romania	2	0	22	13	12	1

Threatened Species

	Number of Threatened Species 2008					
	1 PRIMATES	2 CATS	3 UNGULATES	4 RODENTS	5 BATS	6 DOLPHINS AND WHALES
Russia	–	4	8	7	8	7
Rwanda	2	2	1	5	2	–
Samoa	–	–	–	–	–	1
Saudi Arabia	–	3	4	1	3	–
Senegal	2	3	3	–	1	1
Serbia	–	–	–	2	6	–
Seychelles	–	–	–	–	3	–
Sierra Leone	3	2	4	1	1	–
Singapore	–	1	–	1	1	–
Slovakia	–	–	–	2	5	–
Slovenia	–	–	–	1	6	–
Solomon Islands	–	–	–	4	14	1
Somalia	–	2	10	–	1	–
South Africa	–	3	2	1	3	5
Spain	–	1	–	1	7	6
Sri Lanka	3	3	–	3	–	4
Sudan	1	2	9	–	2	–
Suriname	–	–	1	–	3	3
Swaziland	–	2	1	–	1	–
Sweden	–	–	–	–	3	1
Switzerland	–	–	–	1	3	–
Syria	–	3	5	1	1	–
Tajikistan	–	4	4	1	1	–
Tanzania	4	2	3	2	5	5
Thailand	5	6	8	4	4	2
Togo	2	3	2	–	–	–
Trinidad & Tobago	–	–	–	–	–	–
Tunisia	–	4	6	–	4	1
Turkey	–	4	3	4	6	1
Turkmenistan	–	3	6	3	2	–
Uganda	2	3	1	9	5	–
Ukraine	–	–	2	7	4	1
United Arab Emirates	–	1	4	–	–	1
United Kingdom	–	1	–	–	2	7
United States of America	–	4	–	11	7	9
Uruguay	–	–	1	–	1	3
Uzbekistan	–	4	4	1	2	–
Venezuela	2	–	1	5	5	3
Vietnam	13	5	7	4	3	–
Western Sahara	–	3	5	–	–	1
Yemen	–	2	4	–	1	–
Zambia	–	2	1	2	3	–
Zimbabwe	–	2	1	1	2	–

Number of Threatened Species 2008						
7 REPTILES	8 AMPHIBIANS	9 INVERTEBRATES	10 FISH	11 BIRDS	12 PLANTS	
6	0	29	22	51	7	Russia
0	8	5	9	10	3	Rwanda
1	0	1	8	7	2	Samoa
2	0	2	15	14	3	Saudi Arabia
6	0	0	28	8	7	Senegal
0	0	16	8	10	1	Serbia
10	6	5	14	10	45	Seychelles
3	2	2	16	10	47	Sierra Leone
4	0	1	22	13	54	Singapore
1	0	19	9	7	2	Slovakia
1	2	42	25	3	0	Slovenia
4	2	6	9	20	16	Solomon Islands
2	0	1	25	12	17	Somalia
19	21	152	66	36	73	South Africa
17	5	62	51	15	49	Spain
8	52	52	31	13	280	Sri Lanka
2	0	2	13	13	17	Sudan
6	2	0	21	0	26	Suriname
0	0	0	3	7	11	Swaziland
0	0	13	9	3	3	Sweden
0	1	29	8	2	3	Switzerland
6	0	6	26	11	0	Syria
1	0	2	5	9	14	Tajikistan
5	41	43	137	39	240	Tanzania
22	3	1	50	43	86	Thailand
2	3	0	16	2	10	Togo
5	9	0	21	2	1	Trinidad & Tobago
4	1	7	19	7	0	Tunisia
13	9	12	54	15	3	Turkey
1	0	5	8	15	3	Turkmenistan
0	6	26	54	17	38	Uganda
2	0	14	14	12	1	Ukraine
1	0	2	9	8	0	United Arab Emirates
0	0	10	16	3	13	United Kingdom
32	53	571	166	74	242	United States of America
3	4	1	27	25	1	Uruguay
2	0	1	5	15	15	Uzbekistan
13	69	3	31	26	68	Venezuela
25	15	0	31	38	146	Vietnam
0	0	1	19	1	0	Western Sahara
2	1	6	17	13	159	Yemen
0	1	5	10	10	8	Zambia
0	6	4	3	12	17	Zimbabwe

SOURCES

For sources that are available on the internet, in most cases only the root address has been given. To view the source, it is recommended that the reader type the title of the page or document into Google or another search engine.

PART 1 EXTINCTION IS FOREVER

12–13 EVOLUTION

Darwin C. *On the origin of species by means of natural selection, or The preservation of favoured races in the struggle for life.* first published 1859.
The academic technology web server. daphne.palomar.edu
The Evolution Wing www.ucmp.berkeley.edu/history/evolution.html

14–15 MASS EXTINCTIONS

BBC Learning. www.bbc.co.uk/education
World Resources Institute. www.wri.org

16–17 HOMININS

Institute of Human Origins www.asu.edu/clas/iho/
The TalkOrigins Archive. www.talkorigins.org

18–19 HUMAN ENVIRONMENTAL IMPACT

Arctic sea ice shatters all previous record lows. National Snow and ice Data Center, 2007 October 1. http://nsidc.org/news
Vernon C. The plight of the Great Barrier Reef. 2008 May 23. www.sciencealert.com.au
Global warming: early warning signs. www.climatehotmap.org [accessed March 2008]
Global Footprint Network. www.footprintnetwork.org
Greenemeier L. US protects polar bears under endangered species act. *Scientific American*, 2008 May 14. www.sciam.com
Menzel A et al. Impacts of climate variability, trends and NAO on 20th-century European plant phenology. In: Bronnimann S et al. *Climate variability and extremes during the past 100 years.* Springer Netherlands, 2007. www.springerlink.com
National Geographic, 2004 Sept.
Northwest Territories Environmental and Natural Resources. Caribou forever – our heritage, our responsibility. www.nwtwildlife.com
Human development report 2007/2008. UNDP. pp.102-05. http://hdr.undp.org
Walker G and King D. *The hot topic.* London: Bloomsbury, 2008. pp.39-40.

Part 2 ECOSYSTEMS

22–23 TROPICAL FORESTS

Food and Agricultural Organization. *State of the world's forests 1999.* FAO, 1999. www.fao.org
Forests for life. WWF-UK, 2005. www.panda.org
Bryant D, Nielsen D and Tangley L. *The last frontier forests: ecosystems and economics on the edge.* World Resources Institute, 1997. www.wri.org

Tropical Forest
Protected areas
Tropical forest by region
EarthTrends: The Environmental Portal. www.earthtrends.wri.org
Indonesia: Butler RA. Facts on Borneo. www.mongabay.com/borneo.html [accessed 2008 July 15]
Rainforest loss
FAO. Global forest resources assessment. Downloaded from WDI online.
Rainforest information. www.mongabay.com [accessed 2008 May]

24–25 TEMPERATE FORESTS

Food and Agricultural Organization. *State of the world's forests 1999.* FAO, 1999. www.fao.org
Forests for life. WWF-UK, 2005. www.panda.org
Bryant D, Nielsen D and Tangley L. *The last frontier forests: ecosystems and economics on the edge.* World Resources Institute, 1997. www.wri.org

Temperate Forest
Temperate forest by region
EarthTrends: The Environmental Portal. www.earthtrends.wri.org
China: Liang Chao. Tree-planting paying off as 18% growth recorded. *China Daily,* 2008 June 11. www.chinadaily.com
Russia: Markus F. Illegal logging spreads in Russia. 2001 October 29. http://bbc.co.uk
Wood production
FAO data, accessed via EarthTrends: The Environmental Portal. www.earthtrends.wri.org

26–27 GRASSLANDS

World resources 2000-2001, World Resources Institute, 2000. pp.119–139. www.wri.org
Sauer JR et al. The North American breeding bird survey, results and analysis 1966-98. Version 98.1, USGS Patuxent Wildlife Research Center, 1999.
Grasslands
White R, Murray S, and Rohweder M. *PAGE analysis,* Map 15. World Resources Institute. www.wri.org
Current Land Use: *World resources 2000-2001.* op. cit. p.123.
Asian steppes: *World resources 2000-2001*, op. cit. pp.212–24.
Great Indian bustard: Birding in India and South Asia. www.birding.in [accessed May 2008]

28–29 WETLANDS

Wetlands of International Importance
Wetlands International. www.ramsar.org
International Rivers Network. www.irn.org
WWF. www.panda.org
Financial value: *In the front line. Shoreline protection and other ecosystem services from mangroves and coral reefs.* UNEP, 2006. pp.12-14. www.unep-wcmc.org
Florida mangroves: www.unep-wcmc.org
Water hyacinth: *Biocontrol News and Information*, 2004 Dec, 25(4). www.pestscience.com
Mangrove loss in Thailand
Database on local coping strategies. Mangrove reforestation in southern Thailand. Summary. http://maindb.unfccc.int [accessed 2008 July 15]

30–31 CORAL REEFS

The State of Coral
Wilkinson C. *The status of coral reefs in the world.* WWF, 2004. www.globalissues.org
Veron C. The plight of the Great Barrier Reef. *ScienceAlert,* 2008 May 23. www.sciencealert.com.au
Butler RA. Coral reefs and mangroves have high economic value. 2006 Jan 24. http://news.mongabay.com
Carpenter K, Livingstone S. IUCN report to Coral Reef Symposium, 2008 July 10. Reported by Radford T. One third of reef-building corals face extinction, study shows. *The Guardian,* 2008 July 10. www.guardian.co.uk

32–33 OCEANS

Freiwald A et al. *Cold-water coral reefs.* Cambridge, UK: UNEP-WCMC, 2004.
World Energy Council. *Survey of energy resources 2001.* www.worldenergy.org
Biogenic reefs - cold water corals. Joint Nature Conservation Committee. www.jncc.gov.uk [accessed 2008 July 15]
Boswell R. *Buried treasure.* The American Association of Mechanical Engineers, 2005 Feb. www.memagazine.org
Greimel H. Resource-poor Japan is banking on ice that burns as future fuel source. 2003 July 3. www.climateark.org
NURP research supports conservation of deep-sea corals. 2007 July 16. www.nurp.noaa.gov
New research reveals clear scientific reasons for the bottom trawling to stop. 2006 Nov 15. www.savethehighseas.org
Fisheries organizations in the Atlantic and Indian Oceans fail their end of year report. 2007 Dec 10. www.savethehighseas.org
UN FAO: Meeting ends without agreement on guidelines for high seas deep-sea fisheries. 2008 Feb 10. www.savethehighseas.org

Orange roughy
Lack M., Short K. and Willock A. *Managing risk and uncertainty in deep-sea fisheries: lessons from Orange Roughy*. TRAFFIC Oceania and WWF Australia, 2003. www.wwf.org.au
Polymetallic nodules
Deep-sea bed polymetallic nodule exploration: Development of environmental guidelines. www.isa.org.jm
International Seabed Authority. Marine mineral resources, factsheet. www.isa.org
Workshop on polymetallic nodule mining technology – current status and challenges ahead. 2008 Feb. Background document. Jointly organized by The International Seabed Authority (ISA) and Ministry of Earth Sciences of the Government of India.
Methane hydrates: Japan, US agree to cooperate on methane hydrates. *China View*, 2008 June 7. www.chinaview.cn
Fishing the deep
Giannia M. High seas bottom trawl fisheries and their impacts on the biodiversity of vulnerable deep-sea ecosystems: Options for international action. IUCN, 2004.

PART 3 FRAGILE REGIONS

36–37 THE ARCTIC
Map created by Hugo Ahlenius and reproduced by kind permission of UNEP/GRID-Arendal. Available at: http://maps.grida.no/go/graphic/arctic-conservation-area-caff-topographic-map
Zöckler C and Lysenko I, *Water birds on the edge*. World Conservation Monitoring Centre, 2000.
Cold wars: Russia claims Arctic land. *Geotimes*, 2007 August 1. www.geotimes.org
Protection and Incursion
Protected areas in the Arctic, Arctic transportation routes, The decrease of Arctic sea ice. UNEP/GRID-Arendal. Maps and Graphics Library, 2007. http://maps.grida.no
Arctic National Wildlife Refuge: WWF. Oil and gas in the Arctic. 2008 May 21. www.panda.org
Russian spillage: Russia oil spill. TED case study. 1997 Nov 1. www.american.edu/ted/komi.htm
Red-Breasted Goose
BirdLife International. www.birdlife.org

38–39 THE ANTARCTIC
Reaney P. Antarctic warming killing off fish food. 2004 Nov 4. www.abc.net.au
Davis TH. Tourism threatens Antarctica. Quoting the International Association of Antartic Tour Operators. 2007 June 5. http://travel.timesonline.co.uk
Fogarty D. Race for Antarctic krill a test for green management. Reuters. 2008 May 26.
IUCN Red List. www.redlist.org
Larsen B Ice-Shelf Collapses. 2002 Mar 21. www.nsidc.org
McKenna P. Warming seas threaten Antarctic marine life. 2008 Feb 16. *New Scientist Environment*. http://environment.newscientist.com
National Geographic, 2004 Sept.
The World Factbook 2000. Central Intelligence Agency.
Protected Areas
Secretariat of the Antarctic Treaty. ASPA map and list. www.ats.aq
Southern Ocean Sanctuary: International Whaling Commission. www.iwcoffice.org

40–41 AUSTRALIA
National Geographic, insert on Australia. Washington, 2000 July.
Marine Protected Areas. www.environment.gov.au
Taylor R. Australia to build climate corridor. Reuters. 2007 July 9. http://uk.reuters.com
Macintosh A and Wilkinson D. Environment Protection and Biodiversity Conservation Act. A five-year assessment. The Australia Institute. Discussion paper no. 81. 2005 July. www.tai.org.au
World Heritage Sites
www.unesco.org/whc
Protected areas
Type of protected area
Collaborative Australian Protected Areas Database 2006. www.environment.gov.au
Threatened species
IUCN Red List. www.redlist.org
Gouldian finch: Australian Wildlife Conservancy. www.australianwildlife.org
Orange-bellied parrot: http://www.parks.tas.gov.au/wildlife/birds/obp.html

42–43 CENTRAL AND SOUTH AMERICA
Barreto P et al. *Human pressure on the Brazilian Amazon forests. An overview of the findings and maps.* Mar 2006. World Resources Institute, Imazon. Global Forest Watch. www.globalforestwatch.org
Vidal J. Road to oblivion. *The Guardian*, 2001 June 11. quoting William Laurence of Smithsonian Institute in *Science*, 2001 Jan 19.
WWF. Amazon Region Protected Areas Programme. www.panda.org
WWF. Brazil announces new Amazon protected areas. 2008 May 30. www.panda.org
Protected Land
EarthTrends: The Environmental Portal. http://earthtrends.wri.org
World Heritage Sites: www.unesco.org/whc
Muriqui: The Nature Conservancy. Places we work. The Atlantic forest of Brazil. www.nature.org
Conservation International. Atlantic Forests. www.biodiversityhotspots.org
Unofficial roads: Bellos A. The long road to ruin for the Amazon forest. *The Observer*. 2007 April 15. www.guardian.co.uk
Souza C Jr et al. *The expansion of unofficial roads in the Brazilian Amazon. State of the Amazon*. 2005 May. www.imazon.org.br
Panatal: WWF. Pantanal Program. www.panda.org

44–45 GALAPAGOS
Galapagos Conservation Trust. www.gct.org
Darwin Foundation. www.darwinfoundation.org
Goats: www.galapagos.org/conservation/projectisabela.html
Galapagos penguin: Vargas H, Lougheed C and Snell H. Population size and trends of the Galapagos Penguin Spheniscus mendiculus. *Ibis*, 2005,147. pp. 367-74.

46–47 MADAGASCAR
Protected Areas
www.parcs-madagascar.com/carte.htm
World Heritage Site: http://whc.unesco.org
Threatened animals, Degree of threat, Threatened plant species
IUCN Redlist. www.redlist.org
Madagascar rosy periwinkle: www.mobot.org
Fish eagle, golden lemur: WWF. www.panda.org

Part 4 ENDANGERED ANIMALS AND PLANTS

50–51 PRIMATES
Day P. British biologist warns of threat to monkeys. *Daily Telegraph*, 1999 Oct 31. p.16.
Duke University Primate Center. www.duke.edu/web/primate
University of Manitoba www.umanitoba.ca
Threatened Primates
IUCN Red List. www.redlist.org
Cameroon: Cameroonian gorillas arrive home. 2007 Nov 30. http://news.bbc.co.uk

52–53 CATS
Threatened Cats
IUCN Red List. www.redlist.org
Decline of the tiger
Big Cat Rescue. www.bigcatrescue.org
Bengal tiger: Nitin Sethi. Just 1,411 tigers in India. *Times of India*, 2008 Feb 13. http://timesofindia.indiatimes.com
Florida panther: Florida Panther Net. www.floridaconservation.org/panther

54–55 UNGULATES
Threatened Ungulates
IUCN Red List. www.redlist.org
Arabian oryx: World Heritage Sites. http://whc.unesco.org
Regional strategy for the conservation of Arabian Oryx. *Jordan Times*, 2007 May 30. www.arabenvironment.net
Hirola: Butynski TM. Independent evaluation of hirola antelope (Beatragus hunteri) conservation status and conservation action in

Kenya. Nairobi, 2000, chapter 7. http://coastalforests.tfcg.org
Baby pygmy hog: Durrell Wildlife Conservation Trust. www.durrellwildlife.org
Changing fortunes of the North American bison
Carter D. Future of the buffalo business. National Bison Association. www.bisoncentral.com

56–57 ELEPHANTS AND RHINOS
Distribution of Elephants and Rhinos
IUCN Red List. www.redlist.org
International Rhino Foundation. www.rhinos-irf.org

58–59 BEARS
Kemf E, Wilson A and Servheen C. *Bears in the wild.* WWF, 1999.
Threatened Bears, Threatened Pandas
IUCN Red List. www.redlist.org
Population of threatened bears
WWF. www.panda.org
Brown bear: Hunters kill last female bear native to Pyrenees. 2004 Nov 4. www.timesonline.co.uk
Giant pandas: Elegant S. When pandas go wild. *Time.* 2007 Jan 6. www.time.com
WWF. Panda factsheet. www.panda.org

60–61 RODENTS
Nowak R. *Walker's mammals of the world*, 6th ed. vol II. Baltimore and London: The Johns Hopkins University Press, 1999.
Threatened Rodents
IUCN Red List. www.redlist.org
A Marmot on the Brink
Bryant A, Forestry and historical population dynamics of the endangered Vancouver Island marmot, Contributed talk, American Society of Mammalogists meeting, Durham, New Hampshire, June 2000. www.speciesatrisk.gc.ca
www.marmots.org

62–63 BATS
Threatened Bats
IUCN Red List. www.redlist.org
Rodrigues fruit bat: Factsheet provided by Durrell Wildlife Conservation Trust. www.durrell.org

64–65 DOLPHINS AND WHALES
International Whaling Commission. www.iwcoffice.org
Threatened Dolphins and Whales
IUCN Red List. www.redlist.org
Yangtze River Dolphin: Baiji dolphin previously thought extinct spotted in Yangtze river. *Science Daily,* 2007 Sept 1. www.sciencedaily.com
Great whale populations
International Whaling Commission. www.iwcoffice.org

66–67 REPTILES AND AMPHIBIANS
Disasters take their toll on amphibians. www.climateark.org
Global Amphibian Assessment. www.globalamphibians.org
Threatened Reptiles, Threatened Amphibians
IUCN Red List. www.redlist.org
EarthTrends. The Environmental Portal. http://earthtrends.wri.org
Panamanian golden frog: Last wave of wild golden frog. 2008 Feb 2. http://news.bbc.co.uk
Frog chytrid fungus. www.environment.nsw.gov.au

68–69 INVERTEBRATES
Threatened Invertebrates
IUCN Red List. www.redlist.org
Harlequin ladybird: Hunt for the Harlequin – a UK survey for the world's most invasive ladybird. Natural Environment Research Council, press release, 2005 March 15. www.nerc.ac.uk
Monarch butterflies: Oberhauser K and Townsend Peterson A. Modeling current and future potential wintering distributions of eastern North American monarch butterflies. *Proceedings of the National Academy of Sciences of the United States of America 2003,* 100. pp.14063-68
Threatened crustaceans
IUCN Red List. www.redlist.org

70–71 FISH
Threatened Fish
IUCN Red List. www.redlist.org
Global production
FAO. World fisheriess production by capture and aquaculture 2005. www.fao.org

72–73 PLANTS
Threatened Plants
IUCN Red List. www.redlist.org
Mandrinette: IUCN Red List. www.redlist.org
Bastard quiver tree: Yun L. Saving the bastard quiver tree. IUCN. 2007 July 8. www.conservation.org
Extent of the threat
IUCN Red List. www.redlist.org

Part 5 ENDANGERED BIRDS

76–77 BIRDS
Threatened Birds
IUCN Red List. www.redlist.org
Bald eagle: American Bald Eagle Information. www.baldeagleinfo.com
Tristan albatross, Gough bunting: Vidal J. From stowaway to supersize predator: the mice eating rare seabirds alive. *The Guardian*, 2008 May 20. p.3.
Farmland birds
European Bird Census Council. Trends of common birds in Europe, 2007 update. www.ebcc.info/trends2007.html

78–79 BIRDS OF PREY
Threatened Birds of Prey
IUCN Red List. www.redlist.org
Californian condor in crisis
US Fish & Wildlife Service. 1996 Recovery Plan for the California Condor. http://ecos.fws.gov
CRES. Milestones in Californian Condor Conservation. http://cres.sandiegozoo.org
Decline of Asian white-backed vulture
Prakash V. Bombay Natural History Society. quoted on: www.peregrinefund.org/conserv_vulture.html
O'Connell S. Change in the pecking order. *The Guardian*, 2002 October 24. www.guardian.co.uk
Birdlife International. Hopes soar after vulture chick hatches. 2007 January 9. www.birdlife.org
Great Philippine eagle: WWF. www.panda.org

80–81 PARROTS AND COCKATOOS
Snyder N et al, editors. *Parrots: Status Survey and Conservation Action Plan.* IUCN, 2000.
WWF. www.panda.org
Threatened Parrots and Cockatoos
IUCN Red List. www.redlist.org
Kakapo: Kakapo Recovery Programme. www.kakaporecovery.org.nz
Mauritius parakeet: Birdlife International. www.birdlife.org
Spix's macaw: Al Wabra Wildlife Preservation. Factsheet. http://awwp.alwabra.com
Yellow-eared conure: International Conure Association. www.conure.org
Puerto Rican Amazon parrot
Quist C F et al. Granular cell tumor in an endangered Puerto Rican Amazon parrot. www.vet.uga.edu
Birdlife International facsheet. www.birdlife.org

82–83 SEABIRDS
United Nations General Assembly Resolution 46-215. 1991 Dec 20.
WWF. www.panda.org
Threatened Seabirds
IUCN Red List. www.redlist.org
Madeira petrel, spectacled petrel, short-tailed albatross: www.uct.ac.za
Birdlife International. Factsheet. www.birdlife.org

84–85 MIGRATORY BIRDS
WWF. www.panda.org
The Ramsar Convention on Wetlands. www.ramsar.org
World migratory bird day 2007. www.worldmigratorybirdday.org
Climate change linked to migratory bird decrease. *Science Daily,* 2003 Mar 23. www.sciencedaily.com

Bird migration at mercy of weather patterns. *New Scientist*, 2008 May 17. http://environment/newscientist.com
Americas
National Audubon Society. www.audubon.org
US Forest Service. www.r5.fs.fed.us
WWF www.panda.org
Europe and Africa
BirdLife International. www.birdlife.org
Royal Society for the Protection of Birds. www.rspb.org.uk
BirdWatch Ireland. www.birdwatchireland.ie
Guardian Education. 2000 Mar 28.
Migrating birds know no boundaries. www.birds.org.il
www.wildwatch.com
World Heritage Sites. www.unesco.org
World migratory bird day. www.unep-aewa.org
Asia and Oceania
Sanctuary Asia. www.sanctuaryasia.com
World migratory bird day 2008 focuses on biodiversity. Migratory bird numbers plummeting globally – warning signs of a changing environment. African-Eurasian Waterbird Agreement. 2008 May 8. www.unep-aewa.org
Malta: Malta in about-turn on bird hunting laws. 2006 April 3. www.birdlife.org
Maltese 2008 spring hunting season banned by European Court. 2008 April 25. www.birdlife.org
World Migratory Day: World migratory bird day 2008 focuses on biodiversity. Migratory bird numbers plummeting globally – warning signs of a changing environment. African-Eurasian Waterbird Agreement. 2008 May 8. www.unep-aewa.org

Part 6 ISSUES OF CONSERVATION

88–89 ANIMAL BIODIVERSITY

Groombridge B and Jenkins MD. *World atlas of global biodiversity.* World Conservation Press, 2000.
Animal Biodiversity
World resources 2000-2001. World Resources Institute, 2000, Table BI.3. www.wri.org
Endemic species
EarthTrends: The Environmental Portal. http://earthtrends.wri.org
Total Known Species
Invertebrates: *Global Environmental Outlook 3.* United Nations Environment Programme. www.unep.org
Fish: Froese R and Pauly D, editors. *FishBase 04/2008 version.* www.fishbase.org
Birds: LePage D. Avibase: *The World Bird Database 2004.* Port Rowan, Ontario: Bird Studies Canada. www.bsc-eoc.org
Reptiles: European Molecular Biology Laboratory (EMBL). *The EMBL Reptile Database 2004.* Heidelberg, Germany: EMBL. www.embl-heidelberg.de
Amphibians: Information on amphibian biology and conservation 2004. Berkeley, California: AmphibiaWeb. http://amphibiaweb.org
Mammals: Reeder DM, Helgen KM and Wilson DE. Global trends and biases in new mammal species discoveries. Museum of Texas Tech University Occasional Papers no. 269, 2007 Oct 8.

90–91 PLANT BIODIVERSITY

Groombridge B and Jenkins MD. *World atlas of global biodiversity.* World Conservation Press, 2000.
Plant Biodiveristy
World resources 2000-2001. World Resources Institute, 2000, Table BI.3. www.wri.org
Endemic vascular plant species
EarthTrends: The Environmental Portal. http://earthtrends.wri.org
Number of plant species
UNEP-WCMC

92–93 ECOLOGICAL HOTSPOTS

Conservation Focus
Conservation International. www.biodiversityhotspots.org
WWF Protected Areas for a Living Planet. www.panda.org

94–95 CONSERVING ANIMALS

Conservation Zoos
World Association of Zoos and Aquariums. www.waza.org
Amur leopard: Alta Amur Leopard Conservation. www.amur-leopard.org
Mauritius kestrel: Durrell Wildlife Conservation Trust. www.durrellwildlife.org
Fall and Rise of the Nene
Wildlfowl & Wetlands Trust. www.wwt.org.uk

96–97 CONSERVING PLANTS

Plant Conservation Alliance. www.nps.gov/plants/index.htm
Plant Conservation Institutions
East Asia Botanic Gardens. http://202.127.158.171/eabgn/english/index.htm
Canada BCGI. www.bgci.org/canada/inst_map
Data for USA and Africa from BCGI. www.bgci.org/usa
Data for Europe: The International Plant Exchange Network (IPEN). http://www.bgci.org/abs/ipen
Pacific Agricultural Plant Genetic Resources Network (PAPGREN). www.spc.int
China: www.bgci.org
Golden Pagoda: www.redlist.org
Japanese knotweed: www.invasive.org
Magnolia: www.bgci.org
Svalbard Global Seed Vault: www.bgci.org

98–99 CONSERVING DOMESTIC BREEDS

Threatened Domestic Breeds
FAO Domestic Animal Diversity Information System. www.fao.org/dad-is
Endangered domestic breeds
Commission on Genetic Resources for Food and Agriculture. The state of the world's animal genetic resources for food and agriculture. Rome: FAO, 2007. p.39-40.
Definition of terms
Critical: 100 or fewer breeding females, 5 or fewer breeding males, or a population of under 121 and decreasing, with the percentage of breeding females below 80 percent.
Endangered: 101–1,000 breeding females, 6–20 breeding males, or a population of 81–99 and increasing, with the percentage of breeding females above 80 percent.
Critical maintained/Endangered maintained: critical or endangered populations for which active conservation programs are in place or populations are maintained by commercial companies or research institutions.

100–01 IMPORT TRADE

Major Importers
Data from CITES available via EarthTrends: The Environmental Portal. http://earthtrends.wri.org
African gray: African Grey parrots under threat from pet trade. Planet Ark. 2006 July 5. www.planetark.com

102–03 EXPORT TRADE

Major Exporters
Data from CITES available via EarthTrends: The Environmental Portal. http://earthtrends.wri.org
Sturgeon ban: Caspian Sea states to resume caviar trade. CITES press release. 2002 Mar 6. www.cites.org
Pew Institute for Ocean Science expresses concern over new caviar quotas. 2008 May 27. 222.pewtrusts.org
Ivory: African Elephant ivory sales allowed before renewed ban. Environment News Service. 2007 June 14. www.ens-newswire.com
Tiger parts: Authorities act against tiger poachers in Sumatra. 2008 June 4. www.traffic.org

Part 7 WORLD TABLES

106–13 Table 1: PROTECTED ECOSYSTEMS AND BIODIVERSITY

Column 1: FAOSTAT
Columns 2–3: EarthTrends: The Environmental Portal. http://earthtrends.wri.org
Column 4: www.ramsar.org
Columns 5–9: *World resources 2000-2001* Table BI.3.

114–21 Table 2: THREATENED SPECIES

IUCN Redlist. www.redlist.org

INDEX